THE ARCHAEOLOGY OF SEAFARING IN SMALL-SCALE SOCIETIES

Society and Ecology in Island and Coastal Archaeology

UNIVERSITY PRESS OF FLORIDA

Florida A&M University, Tallahassee
Florida Atlantic University, Boca Raton
Florida Gulf Coast University, Ft. Myers
Florida International University, Miami
Florida State University, Tallahassee
New College of Florida, Sarasota
University of Central Florida, Orlando
University of Florida, Gainesville
University of North Florida, Jacksonville
University of South Florida, Tampa
University of West Florida, Pensacola

The Archaeology of Seafaring in Small-Scale Societies

Negotiating Watery Worlds

Edited by
Alberto García-Piquer,
Mikael Fauvelle,
and Colin Grier

Scott M. Fitzpatrick
and Victor Thompson,
Series Editors

UNIVERSITY PRESS OF FLORIDA

Gainesville/Tallahassee/Tampa/Boca Raton
Pensacola/Orlando/Miami/Jacksonville/Ft. Myers/Sarasota

30 29 28 27 26 25   6 5 4 3 2 1

DOI: https://doi.org/10.5744/9780813079493

Library of Congress Cataloging-in-Publication Data
Names: García-Piquer, Alberto, editor. | Fauvelle, Mikael, editor. | Grier, Colin (Colin Foster), editor.
Title: The archaeology of seafaring in small-scale societies : negotiating watery worlds / edited by Alberto García-Piquer, Mikael Fauvelle, Colin Grier.
Description: Gainesville : University Press of Florida, 2025. | Series: Society and ecology in island and coastal archaeology | Includes bibliographical references and index.
Identifiers: LCCN 2025025136 (print) | LCCN 2025025137 (ebook) | ISBN 9780813079493 (hardback) | ISBN 9780813081274 (paperback) | ISBN 9780813075112 (ebook) | ISBN 9780813074177 (pdf)
Subjects: LCSH: Seafaring life—History. | Underwater archaeology. | Island archaeology. | Voyages and travels—History. | Coastal archaeology.
Classification: LCC G540 .A76 2025 (print) | LCC G540 (ebook) | DDC 910.4/5—dc23/eng/20250615
LC record available at https://lccn.loc.gov/2025025136
LC ebook record available at https://lccn.loc.gov/2025025137

The University Press of Florida is the scholarly publishing agency for the State University System of Florida, comprising Florida A&M University, Florida Atlantic University, Florida Gulf Coast University, Florida International University, Florida State University, New College of Florida, University of Central Florida, University of Florida, University of North Florida, University of South Florida, and University of West Florida.

University Press of Florida
2046 NE Waldo Road
Suite 2100
Gainesville, FL 32609
http://upress.ufl.edu

GPSR EU Authorized Representative: Mare Nostrum Group B.V., Mauritskade 21D, 1091 GC Amsterdam, The Netherlands, gpsr@mare-nostrum.co.uk

Photo courtesy of Jane Ames.

This volume of work is dedicated to the scholarship and memory of Kenneth M. Ames (1945–2019). As is clear from the papers within, Ken's thinking looms large in the theoretical and methodological study of seafaring in coastal societies, particularly those of the Northwest Coast of North America.

His main seafaring paper, "Going by Boat: The Forager-Collector Continuum at Sea" (2002), is akin to an early rock 'n' roll album that inspired everyone who heard it to form their own rock band.

More broadly, Ken always keenly recognized that Northwest Coast hunter-gatherer-fisher societies were something quite different from those on which many archaeological models and interpretations had been built. This view is reflected in his treatment of a wide array of topics that captivated him—the emergence of social inequality, the organization of Northwest Coast house-holds, and the political economies of nonstate societies. On these topics he contributed much to a wider body of archaeological and anthropological theory, influencing and collaborating with archaeologists around the globe.

He lived and worked in the Pacific Northwest of North America, but the global reach and enduring impact of his scholarship attests to his depth of intel-lect and, for those who knew him personally, his depth of character.

CONTENTS

FIGURES

TABLES

FOREWORD

The history of humankind can be crafted along a milieu of narratives. One might be the grand diasporas of smaller groups who left Africa and migrated across Europe and Asia, eventually leading to the expansion of our species into virtually every corner of the globe. Another would certainly include the rise of city-states and civilizations premised on various systems of food production and animal domestication. And we would be remiss in also not recognizing the impacts that technological innovations have had, leading to increasingly efficient and effective ways for exploiting resources. But these are not mutually exclusive events, nor do they capture the complexities involved with various cultural and environmental interludes across time and space that have shaped the evolution of our species, biological or otherwise.

What is evident, however, is that the stories often crafted to explain our place in the world and how we got here neglect a key component—water. It is easy to see why. We are primarily a land-based species, on a planet called Earth no less, and the archaeological record we use to decipher what we have done and where we have gone is largely terrestrial and incomplete in nature. Let us not forget that in the heyday of cultural ecology, *Man the Hunter,* published in 1968 by Richard Lee and Irven DeVore, stressed how big-game hunting was the crux on which human societies developed and thrived. While notably critiqued afterward for its overemphasis on the masculine role within society and its failure to acknowledge the contribution of other types of subsistence strategies, the lingering effects were palpable. Intellectually, it has taken decades to overcome.

This volume—following a steady stream of research over the past half century that demonstrates the importance of viewing coasts not as marginal environments but critical to the course of human history—aims to enhance the optics of deep time in this regard by emphasizing the role of water and the craft developed to cross the aquasphere. Although boats and larger ships have long been recognized as being pivotal historically, particularly during the Age of Exploration and what transpired thereafter, less attention has been paid to the roles that watercraft played in smaller-scale societies. As the editors note, and visible throughout the chapters in this volume, boats were truly transformative, allowing human groups to access new resources and landscapes, interact with others, and adjust boundaries both physically and psychologically. The ability to transport people, goods, ideas, and biota—often more efficiently than over land—was logistically powerful. Humans were no longer terrestrially bound—they were now aquatically advantaged.

This volume clearly illustrates that the strength of understanding how humans have harnessed waterways harks back to traditional thought and action and that these long-distance voyages—which have captured the imagination of many (think: the Polynesian diaspora, Norse sagas, or Magellan's voyage)—owe their legacy to the tendrils of waterways that flow through terrestrial habitats. It was the human connection between land and water—using wood and stone and massive amounts of labor to support—that led our species to dissolve the terrestrial tethers that bound us to land.

The ways in which watercraft have moved us both literally and figuratively are the precipices on which we have reached beyond terra firma. Rafts, boats, ships, and the multitude of configurations that humans have created over the eons are a testament to our ingenuity, survival, and curiosity. Their exquisiteness has inspired people and places and have been catalysts for seeking other unknowns beyond both land and the sea's surface. And this tome, with its rich menagerie of cultural traditions and stories, is one that will teach and inspire the next generation of scholars intent on understanding the past and origins of our aquatic lifeways.

Scott M. Fitzpatrick
Series Editor

ACKNOWLEDGMENTS

We would like to thank the Per Anders and Maibrit Westrin Foundation for providing funds for the proofreading and indexing of this book. We would also like to thank the Lund University Library and the Joint Faculties of Humanities and Theology at Lund University for providing funds that enabled us to publish this volume as an open access book. Mikael Fauvelle would like to thank the Riksbankens Jubileumsfond (M21-0018) and the Marcus och Amalia Wallenbergs Minnesfond (2022.0108), both of which provided funds to projects that helped shape his thinking regarding ancient seafaring. Finally, Alberto García-Piquer would like to acknowledge the funding from the European Union's Horizon Europe research and innovation program under the Marie Sklodowska-Curie grant agreement No. 101066298, which provided him the opportunity to encounter a whole new watery world and its captivating history.

1

Why Boats Matter

Breaking Down Terrestrial Bias in Archaeology

MIKAEL FAUVELLE, ALBERTO GARCÍA-PIQUER,
AND COLIN GRIER

When archaeologists tell stories about the past, the stories usually take place on dry land. Archaeological narratives for important events ranging from early human dispersals from Africa to the formation of archaic states have all traditionally focused on terrestrial processes. This terrestrial bias permeates much of what we do, from how we conduct our fieldwork to the theory we use to explain social change. Many of the most important events and transformations of our past, however, did not happen on dry land but were instead contingent on our ancestors' capacity to take to water. The colonization of new continents, coastal and riverine resource intensification, sedentism, and the establishment of long-distance interaction networks, for example, were all facilitated by using boats to access new resources and traverse open waters. This volume seeks to upend the terrestrial bias that runs through much of archaeology by exploring the diverse strategies that ancient people used to negotiate the watery worlds with which they were entangled. In this opening chapter, we make the case for the importance of boats as transformative technological and social innovations that made much of human history possible and structured how it unfolded. Following in the canoe-wake of Kenneth M. Ames (2002: 47), we argue that boats fundamentally matter for anthropological theory as they profoundly structured the relationship between humankind and our surroundings.

Recently there have been numerous calls to rethink the traditional ways in which archaeologists have described the past, yet few of these voices have significantly challenged the overarching terrestrial focus of most archaeological narratives. Graeber and Wengrow (2021), for example, have questioned traditional archaeological approaches to understanding the origins of social inequality, arguing that we should see inspiration in the great variety and complexity of social formations that existed in the past. Virtually all of the examples used in their book, however, from the Wendat Confederacy to Neolithic Ukraine, were primarily land-based cultures. Even when focusing on classic seafaring peoples such as those of the Northwest coast and California, boats are rarely centered in the conversation. Likewise, Scott (2017) has argued for a rethinking of the Neolithic Transition to acknowledge that many hunting and gathering societies took thousands of years to fully adopt new agricultural practices. While Scott does discuss the importance of wetland environments and the aquatic resources they contain, he does not delve into the numerous coastal and island societies that had knowledge of domesticates but chose not to use them, sometimes persisting in fisher-forager lifeways for millennia after their neighbors had become obligate agriculturalists. A recent review of the "Grand Challenges" archaeology can address in the twenty-first century overlooks boats and maritime societies in advancing the 25 archaeological focus areas the authors chose to highlight (Kintigh et al. 2014). Clearly, terrestrial bias is alive and strong in contemporary archaeology.

Unfortunately for the current grand narratives, it is increasingly apparent that much of human history happened on or near the water. Perhaps the most obvious example of the importance of watercraft in the deep past is their use to carry our ancestors to virtually every corner of the globe. We now know that some form of watercraft transported early seafarers to Australia by 65,000 years ago (Clarkson et al. 2017) and that boats of some description were central to the coastal migrations that brought the first humans to the Americas (Davis and Madsen 2020; Erlandson et al. 2007). Early Paleolithic tool assemblages on the island of Crete indicate that even Neanderthals may have taken to the sea in watercraft (Ferentinos et al. 2012; Strasser et al. 2011). In later periods, boats and marine resources are known to have played a central role in the formation of ancient chiefdoms and states in regions ranging from Scandinavia to Hawaii (Kirch 2010; Ling et al. 2018). Conversely, they can also act as mechanisms to break bottlenecks through which increased centralization can typically be established in the political economies of small-scale societies (Furholt et al. 2020). Many of prehistory's most important transformations and migrations, such as the spread of Bell Beaker people and their Chalcolithic technological traditions across the Atlantic facade (Vander Linden 2016) or the migrations of Polynesian peoples across the Pacific (Spriggs 2011; Wilmshurst et al. 2011), were clearly facilitated by seaworthy vessels. Boats mattered for some of the most important events and processes that shaped our shared past.

The importance of boats has been articulated by several previous works, many of which have focused on the logistical capacities of watercraft. In a seminal paper cited many times throughout this book, Ames (2002) examined the logistical capacities of different boat types used by Indigenous peoples of the Pacific Northwest coast of North America. He argued that the use of boats fundamentally changed the relationship between people and their environment. The study of these relationships requires different theoretical approaches that more adequately capture the mobility strategies in aquatic hunter-gatherer societies. While the concern with human–environmental relationships is an area overdue for a broad rethink of the type called for by Ames, its scope should extend beyond the spatial and logistical and into the social and political. For example, the connection between the use of boats and the development of political complexity has been discussed by Arnold (1995), who argued that the transportation capacities of boats allowed elites to expand and control trade in important resources in ancient southern California. The rethink can also tackle the phenomenological and the cognitive dimensions of life in boats. In Scandinavia, Westerdahl (1992) emphasized the importance of boats in the cognitive landscape of peoples inhabiting maritime landscapes, suggesting that we cannot understand maritime societies without accounting for the phenomenological dimension of life and the experience of passing through seascapes in watercraft. The work of these scholars and many others (Clark 2013; Des Lauriers 2005; Erlandson and Braje 2022; Fauvelle 2011; Fitzhugh and Luukkanen 2019; Fitzpatrick 2013; Wright 2014) has led to a gradual recognition of the great importance of studying the role of watercraft in the political, economic, and social life of small-scale societies.

Yet there remains much to be done, and the terrestrial focus and bias we have recounted hinders a more transformative development of new theory. Recognition of the importance of boats in ancient societies has been hampered by the fact that until the late twentieth century, maritime lifeways were underexamined in archaeological literature. This was especially the case for scholars studying hunter-gatherer societies, where it was assumed that most ancient lifeways would parallel those of contemporary and ethnographically studied mobile groups such as the !Kung or Hadza. Maritime environments were seen as marginal, unable to support the same population levels as seen in terrestrial regions where ancient people could hunt game and collect nuts, seeds, and tubers (Cohen 1977; Osborn 1977). Pathways to complexity were likewise seen as terrestrially focused, with agriculture seen as the primary route to sedentism, population growth, and institutionalized political hierarchy (e.g., Johnson and Earle 1987). Marine adaptations, if discussed, were viewed as a late development that occurred in tandem with the "broad spectrum revolution" and the beginnings of neolithization (Flannery 1973). Most maritime archaeology during the twentieth century was therefore focused on the archaeology of the large sailing ships used by states and empires, with small-scale boats and marine adaptations largely left to the sidelines of archaeological inquiry. Exceptions to this focus on island regions far from any landmass. The peopling of remote Pacific islands has long held the interest of archaeologists due to the proficiency and skill of seafaring that would have been required of and fueled by the Indigenous seafaring prowess observed by early European sailors and ethnographers across the Pacific Ocean. Not surprisingly, the earliest attempts to model seafaring using computers were in Polynesia, extending back to the 1970s (Finney 1977; Irwin et al. 1990).

In the case of maritime contexts within larger mainland areas, the importance of maritime resources and coastal regions in the development of social complexity became increasingly recognized toward the end of the twentieth century. Working in Peru during the 1970s, Moseley (1975) found evidence that the earliest complex societies of Andean civilization were focused more on shellfish collection and maritime resources than early domesticates. Other case studies from around the world including Japan, the Pacific Northwest coast of North America, southern California, and Florida showed how fisher-forager societies could develop highly complex sedentary societies through mobilizing the rich resources available in maritime environments (Ames 1994; Arnold 1996; Gill et al. 2019; Habu 2002; Sampson 2023; Sassaman 2004; Widmer 1988). As was recently emphasized by Jeanne Arnold and colleagues (2016), however, these works have still not succeeded in changing the dominant terrestrial narrative of our past, which still focuses on agricultural domestication and neolithization as the primary historical trajectories taken by our ancestors. The field of maritime archaeology also continues to focus on sailing ships and modern shipwrecks, with few comprehensive or comparative studies focused on the archaeology of boat types used by hunter-gatherer groups and other small-scale societies (Ames 2002; cf. Arnold 1995; Ohtsuka 1999).

One reason watercraft may not have received as much attention in the literature on small-scale societies as terrestrial technologies of similar importance is that they can be very hard to identify in the archaeological record. Compared to features such as plowed fields, roads, or walls, watercraft leave very little in the way of material remains for archaeologists to study. This is because they are mobile and perishable by nature and are often lost in high-energy coastal and marine environments. Sea-level changes following the end of glacial times also obscure the archaeological visibility of seafaring, as in many parts of the world the coastlines that were likely used to explore and live on are now underwater— and so, potentially, are the ancient crafts.

Some finds of early boats, however, do exist, usually in the unique and limited contexts of waterlogged sites. Dugout boats, such as the almost 10,000-year-old

Table 1.1. Transformative Properties of Watercraft

Process of Change	Relevant Chapters
Provide access to new resources	4, Schulz Paulsson
Allow for exploration and colonization of new regions	5, Des Lauriers and García-Des Lauriers
Facilitate the transportation of goods and people	2, García-Piquer 6, Aguilera et al. 11, Rivera Prince
Underwrite settlement networks and mobility strategies	2, García-Piquer 3, Rorabaugh 5, Des Lauriers and García-Des Lauriers 6, Aguilera et al.
Shape social relationships and political strategies	7, Thompson 8, Fauvelle and Jordan
Generate ontologies of seafaring	4, Schulz Paulsson 9, Smith 10, Whitridge 11, Rivera Prince 12, Jarrett

Pesse logboat (Lanting 2000), are particularly prone to preservation, while spectacular finds of sewn-plank boats such as the Dover Bronze Age boat (Clark 2013) or the Early Iron Age Hjortspring boat (Crumlin-Pedersen and Trakadas 2003) were deposited in low-oxygen environments that preserved them until modern times. Lakes also create particularly good conditions for the preservation of watercrafts, as evidenced by the discovery of 101 prehistoric dugout canoes, the oldest dating back to 5,000 years ago, on the shore of Newnans Lake, Florida (Ruhl and Purdy 2005). Many early boat fragments have also been found in excavations worldwide and can yield great amounts of information with careful scientific analysis (Gamble 2002; Irwin et al. 2017; Lira et al. 2015; Westerdahl 1985). Indirect methods, such as studies of boatmaking tools (Gamble 2002; Kahn et al. 2022), boatbuilding sites (Ling et al. 2024), raw material sourcing (e.g., overseas movement of obsidian artifacts [Ikeya 2015]), or the navigational capacities of different boat types (Fauvelle and Montenegro 2024), can also help us study boat innovation in the absence of direct boat finds.

Difficulties identifying boats in the archaeological record should in no way distract us from their importance for human history. On the contrary, it makes the need to find robust ways to identify and discuss the impacts of ancient boats even more crucial. The chapters in this book adopt multifaceted and interdisciplinary approaches to study how ancient boats were used and conceived of in a variety of different contexts. Through filling in the gaps in our holistic understanding of the impacts of ancient boats, we can better understand how they shaped the maritime societies that used them. Such approaches are critical as boats were one of the most impactful technological innovations in the history of our species. Together with other major innovations such as fire, writing, money, and the internet, boats expanded the capacities of human societies in ways that could not have been fully imagined or intended by their original users. By allowing for new capacities and opening doors to new ways of living and being in the world, the invention and evolution of boats and watercraft produced rapid and transformative changes in human societies (see Table 1.1).

The Transformative Capacities of Watercraft

One key transformative quality of watercraft is the access they allow to watery worlds that would otherwise be impossible for humans to reach. Near-shore aquatic resources have long been important to humans and our ancestors, with archaeological evidence of aquatic resource exploitation going back at least 165,000 years (Marean et al. 2007). Our closest nonhuman relatives, the Neanderthals, are also known to have made extensive use of marine resources, with faunal material from Middle Paleolithic sites such as Figueira Brava indicating diets rich in shellfish, crustaceans, and fish (Zilhão et al. 2020). While modern humans and our ancestors have always had the capacity to collect near-shore resources, the use of watercrafts would have changed the game for ancient marine subsistence. Near-shore resources accessible by simply getting your feet wet can be exploited in new ways, and going out to sea allows for access to resources that are impossible to access from the shore. Deep-sea pelagic fish, for example, are most readily hunted by boat and were prized by cultures worldwide (Bernard 2004). The hunting of pinnipeds and cetaceans would also have been possible most easily by using maneuverable boats, which is attested to by rock art dating from Late Mesolithic and Neolithic times (Lee and Robineau 2004; see also Schulz Paulsson, this volume: Chapter 4). In addition to providing access to many food resources, hunting these species by boat is also often seen as a prestigious activity and would have brought social rewards above and beyond the caloric returns from meat and oils.

In addition to opening new worlds for human subsistence exploitation, watercraft allowed new possibilities for our ancestors to explore and access the worlds around them. Watercraft would have been necessary for early pioneers to reach the continent of Australia (Clarkson et al. 2017) as well as to colonize vast island regions of the Pacific (Montenegro et al. 2006, 2014). Early colonizers of the Americas would also have relied on boats to hop between islands and ice-free sections of coastal North America, following perhaps the so-proposed kelp highway (see Erlandson et al. 2007; Lesnek et al. 2018). The new worlds accessed through boats would also have included untold numbers of islands throughout the world that are adjacent to mainland areas yet difficult to reach without the assistance of watercraft. These islands were often rich in resources, and their habitation and use allowed for the flourishing of unique maritime-focused societies throughout the world (Erlandson and Fitzpatrick 2006). Boats therefore allowed for the formation of entire communities and lifeways that would not have been possible without their use (Des Lauriers and García-Des Lauriers, this volume: Chapter 5).

Watercraft also allow us to move goods across the landscape at scales that are impossible to achieve overland. By loading goods onto boats and rafts and moving them over the water, we can transport masses that would be inconceivable to transport over land. Even today, the longest freight trains are incapable of transporting as much cargo as is possible with a commercial container ship. Using boats, both ancient and modern societies are able to overcome what the historian Fernand Braudel (1985) described as the "tyranny of distance" to support large territorial states and support major population centers. As described by Algaze (2021), water transportation is on average around eight times more efficient than land-based alternatives, a pattern he argues has held from ancient Mesopotamia to nineteenth-century Europe. The location of many major capital cities near rivers and coasts is a clear testament to the power of water-based transport in moving both goods and people to central places.

The logistical powers of watercraft would have been just as critical for small-scale societies as they are for cities, states, and empires. Mobility is a key element of the hunter-gatherer lifeway as hunters and foragers need to move across the landscape to be able to access variable resources (Kelly 2013). The use of boats completely changes the dynamics of hunter-gatherer mobility by allowing for longer voyages, the ability to move vastly greater quantities of food, and the ability to connect settlements across maritime seascapes (Ames 2002; see also García-Piquer, this volume: Chapter 2). Traveling by boat would mean that larger animals, including sea mammals such as pinnipeds and whales, could be transported back to a central camp instead of being butchered near hunting locations, eliminating the need for certain types of logistic camps (Ames 2002). The long range of fisher-forager hunting trips can be seen in the isotopic signatures of the animals consumed, which can come from up to 30 km distant from village sites (Boethius et al. 2022). The ability to move large amounts of resources to a central place also reduces the need for residential mobility, leading to a tendency for maritime hunter-gatherers to live in sedentary communities (Ames 2002; Boethius 2017; Hayden 1994; Kelly 2013). Increased sedentism, combined with greater logistical capacities, would have changed the playing field for boat-using hunter-gatherer societies.

Watercraft also have the capacity to transform social relationships and alter the course of social trajectories. Extensive social networks and novel types of networks can be established and maintained using the mobility offered by watercraft (see chapters by Thompson [7] and Fauvelle and Jordan [8], this volume). Maritime hunter-gatherers often live in societies with higher degrees of hierarchy and economic stratification. The transportation capacities of boats allow for entrepreneurial individuals to monopolize trade networks, leading to increasing concentrations of wealth in the hands of an economic and political elite (Arnold 1995; Fauvelle et al. 2024). The large amounts of labor, specialized knowledge, and high-value materials needed to build boats can also lead to positive feedback mechanisms that empower boat owners and lead to political centralization (Fauvelle and Jordan, this volume: Chapter 8). Control over boat financing, coupled with increased capacities for trading and raiding, can lead to specialized political-economic formations that have recently been described as a maritime mode of production (Ling et al. 2018), and can lead to parallels in political histories for small-scale maritime societies in many different world regions (Fauvelle and Ling 2024; Hudson 2022). Conversely, boats can also offer opportunities for those fleeing authoritarian structures to vote with their paddles and more readily move to new locations outside of chiefly control (Furholt et al. 2020). The use of boats, therefore, can have transformative outcomes for maritime societies that parallel the historical consequences of domestication and adaptation of agriculture for more terrestrially focused cultures.

Boats also shape the worldviews of people who use them. Seeing the world from the water provides fundamentally different insights than terrestrial perspectives and can lead to unique cosmological and phenomenological perspectives. This includes mental geographies of landscapes and seascapes in which boat-using people have different understandings of places, routes, and space than their terrestrially focused neighbors (see chapters by Rorabaugh [3] and Jarrett [12], this volume; Westerdahl 1992). It can also involve ontological perspectives in which boats are seen as important beings on par with animal and human actors (see chapters by Smith [9] and Whitridge [10], this volume). Ethnographic and historical studies of boat use can complement archaeological information to better understand how the use of boats affected not only economic and political structures but also the broader ways boats shaped human worlds (see Aguilera et al., this volume: Chapter 6). Such knowledge can also contribute to engagement with contemporary Indigenous communities for whom boats continue to be a central component of daily life.

Notes on Terminology

Watercraft come in many different forms, and the examples of watercraft described in this book range from relatively simple dugout canoes to large clinker-built sailing ships. Watercraft provides a generic term for the vessels people used to take to water and encompasses types of craft ranging from makeshift rafts to modern ships. Canoes and kayaks were some of the first watercraft used to take to the sea and are defined as narrow craft that are pointed on both ends and generally propelled by paddling. Boats and ships are often larger than canoes, with ships being relationally classified as larger than boats. As terminology surrounding watercraft use can vary considerably around the world, we have let each chapter author work with the terminology that best suits their discussion of the different watercraft that are featured in each of the case studies in this book.

There are also many different ways to describe how people move across the water. The term "seagoing" is used in many chapters in this book to describe the regular and widespread transport, fishing, and hunting undertaken by many maritime peoples around the world. Seafaring is a similar term that can also imply more frequent trips on the open ocean and a deeper and broader engagement with life on the water. Voyaging generally refers to longer and more directed trips—for example, the colonization voyages undertaken by ancient Polynesians or the long-distance voyages to acquire continental copper undertaken by Bronze Age Scandinavians. While these different types of activities would have been different, they would all have required the use of watercraft and a deep knowledge of how to negotiate the sea.

Watercraft are transformative as they shape the way in which people living in island and coastal environments negotiate (that is, navigate and engage with) the watery worlds that surround them. We see these watery worlds as encompassing both the physical seascapes that surround oceangoing societies and the spiritual and cosmological interpretations of those worlds envisioned by such people. Invariably, the physical and cosmological elements of seafaring are heavily intertwined, as we discuss more fully in the concluding chapter. In this book we focus on island and coastal communities as these are areas in which water and seagoing pose a set of unique challenges and possibilities. Societies in riverine regions also often used watercraft in comparable ways to those in coastal areas but differ in that their worlds were circumscribed by dry land. For people living on the sea, it was often water that defined both the boundaries and possibilities of their world, making it critical for us to understand how these watery worlds were negotiated, navigated, and understood.

Organization of Book

The transformative power of boats makes it critical that we understand their impact on early and small-scale societies. The logistical, political, economic, and ontological impacts of watercraft on the societies that used them were enormous, making boats one of the most pivotal technological innovations in human history. As these transformative properties developed in both parallel and disparate ways in different societies around the world, it is important that we develop an appreciation for the diversity of practices and adopt comparative approaches to understand their impact, a key dynamic we return to in the concluding chapter. Pursuing both themes, this volume is an explicit effort to bring together scholars working in North and South America, Western and Northern Europe, and East Asia and drawing on a range of perspectives based on archaeological evidence, historical records, and ethnographic studies to describe the ways small-scale communities used boats to negotiate the watery worlds that surround them, and the ways those practices have been shaped by the commonalities of life at sea around the globe.

For this volume we have by design limited our scope to mostly saltwater contexts and to what can be coarsely referred to as small-scale societies (contra Bird et al. 2019, we use "small scale" as a measure of sociopolitical integration rather than the spatial extent of interactions). This leaves by the wayside the well-studied arena of oceangoing shipping that has been engaged in by pre-industrial and later states as a means to fuel their economic vitality and often their expansion. It also leaves under-addressed the many major river systems that were critical highways through terrestrial continents. As is clear from global histories, the "rivers as highways" is a critical theme in the trajectories of social change on planet Earth. River mouths, as the nexus between inland areas, river transportation corridors, and ocean expanses, played a critical role in how both large- and small-scale societies emerged, changed, and persisted (Ames 2002; Wengrow 2018).

The chapters in this book are organized thematically and reflect the main components of what we argue represents an integrated and expansive approach to theorizing boats and watercraft—(1) movement and logistics, (2) sociopolitical organization and change, and (3) ontology and phenomenology. The first section of the book includes chapters that examine the contribution of watercraft to facilitating logistics and movement in maritime and aquatic societies. Here chapters focus on how we can model the movements of boats across the seascape, map the settlement of ancient people on islands and along coastlines, and understand how boats shape the capacities of people who used them.

The opening chapter of this section is authored by Alberto García-Piquer, who uses an agent-based modeling perspective to test models of settlement organization in Western Patagonia. García-Piquer uses the results of his model to argue that the creation of cultural seascapes is best seen as a negotiation between geography, technology, and the organizational strategies used by Indigenous people. Agent-based modeling is also used in a chapter by Adam Rorabaugh to explore trip lengths and settlement dynamics in the Coast Salish region of the Pacific Northwest coast, arguing that digital storytelling can complement archaeological narratives to build a richer understanding of maritime transport in the past. In the following chapter, Bettina Schulz Paulsson examines the rock art record of Atlantic Europe and South America to build an argument for the antiquity of whale hunting by Megalithic societies in Brittany. A chapter by Matthew Des Lauriers and Claudia García-Des Lauriers examines the long-term history of maritime mobility in Baja California and discusses how access to the sea has shaped the cultural geography of the region. Finally, a chapter contributed by Nelson Aguilera and colleagues uses historical documents to trace the development and use of different boat types in southern Patagonia, showing how the connection between landscape, ecology, and the needs of Indigenous communities shaped the social history of maritime navigation in that region.

The middle section of the volume contains two chapters that focus on the ways boats shape and transform sociopolitical processes in island, coastal, and riverine societies. The first chapter of this section, contributed by Victor D. Thompson, examines how communities in ancient Florida built canals to modify the landscape in ways that facilitated aquatic transportation. Thompson argues that canal construction required collective and cooperative institutions, and he explores how these construction projects shaped the cultural history of the region. Finally, a chapter by Mikael Fauvelle and Peter Jordan compares the use of a specific type of watercraft, the sewn-plank canoe, between the Ainu region of Japan, Bronze Age Scandinavia, and pre-colonial California. In all three cases, they argue that the innovation of the plank canoe led to increases in regional economic integration and the development of political hierarchies, showing the deep connection between watercraft use and sociopolitical organization.

The final section includes chapters taking an ontological perspective on boats and the human–boat relationship. Chapters by Erin Smith (9) and Peter Whitridge (10) examine the relationship between human beings and the boats they used to navigate the watery worlds that surrounded them. Smith focuses on the Pacific Northwest coast, where she explores boat–human relations within

Indigenous ontologies. Likewise, Whitridge describes the ontological relationships between humans, boats, and caribou for Inuit hunters in the Eastern Arctic. The entangled connection between humans, boats, and the sea is also explored from a bioarchaeological perspective in a chapter by Jordi A. Rivera Prince, who examines how the use of boats can shape and leave biological markers on the bodies of fishing peoples. Rounding out this section, Greer Jarrett explores mental geographies of seafarers on the west coast of Norway, using modern experimental seafaring as a window with which to examine the ways in which Viking navigators would have navigated through their watery world.

These chapters are bookended by introductory and concluding chapters by the volume editors in which, starting from the recognition that watercraft represent one of the most pivotal technological innovations in our past, we outline a holistic and integrated approach to the study of seafaring. The goal of this volume is to offer chapters that are exemplary of the various aspects of the study of seafaring we wish to highlight and, in the concluding chapter, provide an integrative approach to taking the project of seafaring forward. In our conclusion chapter we also highlight key areas for future theory-building regarding the role of watercraft in shaping the human experience in an explicit effort to provide a path forward that goes beyond the terrestrial bias to theory we have highlighted here. Throughout the volume, we emphasize the importance of boats for understanding small-scale societies and argue that we cannot adequately chart the history of our species without accounting for the central importance of watercraft.

References Cited

Algaze, Guillermo. 2021. The Tyranny of Friction. *Aegyptus* 42: 73–92.

Ames, K. M. 2002. Going by Boat: The Forager-Collector Continuum at Sea. In *Beyond Foraging and Collecting: Evolutionary Change in Hunter-Gatherer Settlement Systems,* edited by Ben Fitzhugh and Junko Habu, pp. 19–52. Kluwer Academic/Plenum Publishers, New York.

Ames, Kenneth M. 1994. The Northwest Coast: Complex Hunter-Gatherers, Ecology, and Social Evolution. *Annual Review of Anthropology* 23: 209–229.

Arnold, J. E. 1995. Transportation Innovation and Social Complexity Among Maritime Hunter Gatherer Societies. *American Anthropologist* 97(4): 733–747.

Arnold, J. E. 1996. The Archaeology of Complex Hunter-Gatherers. *Journal of Archaeological Method and Theory* 3(1): 77–126.

Arnold, Jeanne E., Scott Sunell, Benjamin T. Nigra, Katelyn J. Bishop, Terrah Jones, and Jacob Bongers. 2016. Entrenched Disbelief: Complex Hunter-Gatherers and the Case for Inclusive Cultural Evolutionary Thinking. *Journal of Archaeological Method and Theory* 23(2): 448–499.

Bernard, Julienne. 2004. Status and the Swordfish: The Origins of Large-Species Fishing among the Chumash. In *Foundations of Chumash Complexity*, edited by Jeanne E. Arnold, pp. 25–52. Cotsen Institute of Archaeology Press at UCLA, Los Angeles.

Bird, Douglas W., Rebecca Bliege Bird, Brian F. Codding, and David W. Zeanah. 2019. Variability in the Organization and Size of Hunter-Gatherer Groups: Foragers Do Not Live in Small-Scale Societies. *Journal of Human Evolution* 131: 96–108.

Boethius, Adam. 2017. Signals of Sedentism: Faunal Exploitation as Evidence of a Delayed-Return Economy at Norje Sunnansund, an Early Mesolithic Site in South-Eastern Sweden. *Quaternary Science Reviews* 162: 145–168.

Boethius, Adam, Melanie Kielman-Schmitt, and Harry K. Robson. 2022. Mesolithic Scandinavian Foraging Patterns and Hunting Grounds Targeted Through Laser Ablation Derived 87Sr/86Sr Ratios at the Early-Mid Holocene Site of Huseby Klev on the West Coast of Sweden. *Quaternary Science Reviews* 293: 107697. http://doi.org/10.1016/j.quascirev.2022.107697.

Braudel, Fernand. 1985. *The Structures of Everyday Life: 1. Civilization and Capitalism*. Fontana Press, London.

Clark, Peter. 2013. *The Dover Bronze Age Boat*. English Heritage, London.

Clarkson, Chris, Zenobia Jacobs, Ben Marwick, Richard Fullagar, Lynley Wallis, Mike Smith, Richard G. Roberts, Elspeth Hayes, Kelsey Lowe, and Xavier Carah. 2017. Human Occupation of Northern Australia by 65,000 Years Ago. *Nature* 547(7663): 306–310.

Cohen, Mark. 1977. *The Food Crisis in Prehistory: Overpopulation and the Origins of Agriculture*. Yale University Press, New Haven, Connecticut.

Crumlin-Pedersen, Ole, and Athena Trakadas (editors). 2003. *Hjortspring: A Pre-Roman Iron-Age Warship in Context*. Viking Ship Museum, Roskilde, Denmark.

Davis, Loren G., and David B. Madsen. 2020. The Coastal Migration Theory: Formulation and Testable Hypotheses. *Quaternary Science Reviews* 249: 106605.

Des Lauriers, M. R. 2005. The Watercraft of Isla Cedros, Baja California: Variability and Capabilities of Indigenous Seafaring Technology along the Pacific Coast of North America. *American Antiquity* 70(2): 342–360.

Erlandson, Jon M., and Todd J. Braje. 2022. Boats, Seafaring, and the Colonization of the Americas and California Channel Islands: A Response to Cassidy (2021). *California Archaeology* 14(2): 159–167.

Erlandson, Jon M., and Scott M. Fitzpatrick. 2006. Oceans, Islands, and Coasts: Current Perspectives on the Role of the Sea in Human Prehistory. *Journal of Island & Coastal Archaeology* 1(1): 5–32.

Erlandson, Jon M., Michael H. Graham, Bruce J. Bourque, Debra Corbett, James A. Estes, and Robert S. Steneck. 2007. The Kelp Highway Hypothesis: Marine Ecology, the Coastal Migration Theory, and the Peopling of the Americas. *Journal of Island and Coastal Archaeology* 2(2): 161–174. https://doi.org/10.1080/15564890701628612.

Fauvelle, Mikael. 2011. Mobile Mounds: Asymmetrical Exchange and the Role of the Tomol in the Development of Chumash Complexity. *California Archaeology* 3(2): 141–158.

Fauvelle, Mikael, and Johan Ling. 2025. Larger Boats, Longer Voyages, and Powerful Leaders: Comparing Maritime Modes of Production in Scandinavia and California. In *Maritime Encounters I: Presenting Counterpoints to the Dominant Terrestrial Narrative of European Prehistory,* edited by John T. Koch, Mikael Fauvelle, Barry Cunliffe, and Johan Ling. Oxbow Books, Oxford.

Fauvelle, Mikael, and Álvaro Montenegro. 2024. Do Stormy Seas Lead to Better Boats? Exploring the Origins of the Southern Californian Plank Canoe Through Ocean Voyage Modeling. *Journal of Island & Coastal Archaeology* 20(3): 1–21. https://doi.org/10.1080/15564894.2024.2311107.

Fauvelle, Mikael, Sasaki Shiro, and Jordan Peter. 2024. Maritime Technologies and Coastal Identities: Seafaring and Social Complexity in Indigenous California and Hokkaido. *Indigenous Studies and Cultural Diversity* 1(2): 30–52.

Ferentinos, George, Maria Gkioni, Maria Geraga, and George Papatheodorou. 2012. Early Seafaring Activity in the Southern Ionian Islands, Mediterranean Sea. *Journal of Archaeological Science* 39(7): 2167–2176.

Finney, Ben R. 1977. Voyaging Canoes and the Settlement of Polynesia: Sailing Trials with Reconstructed Double Canoes Show That Intentional Settlement of Polynesia Was Possible. *Science* 196(4296): 1277–1285.

Fitzhugh, W. W., and H. T. Luukkanen. 2019. The Indigenous Watercraft of Northern Eurasia. *Вестник СПбГУ. История* 64(2): 475.

Fitzpatrick, Scott M. 2013. Seafaring Capabilities in the Pre-Columbian Caribbean. *Journal of Maritime Archaeology* 8: 101–138.

Flannery, K. V. 1973. The Origins of Agriculture. *Annual Review of Anthropology* 2: 271–310.

Furholt, Martin, Colin Grier, Matthew Spriggs, and Timothy Earle. 2020. Political Economy in the Archaeology of Emergent Complexity: A Synthesis of Bottom-Up and Top-Down Approaches. *Journal of Archaeological Method and Theory* 27: 1–35.

Gamble, L. H. 2002. Archaeological Evidence for the Origin of the Plank Canoe in North America. *American Antiquity* 67(2): 301–315.

Gill, Kristina M., Mikael Fauvelle, and Jon M. Erlandson. 2019. *An Archaeology of Abundance: Reevaluating the Marginality of California's Islands.* University Press of Florida, Gainesville.

Graeber, David, and David Wengrow. 2021. *The Dawn of Everything: A New History of Humanity.* Penguin, London.

Habu, Junko. 2002. Jomon Collectors and Foragers. In *Beyond Foraging and Collecting,* edited by Ben Fitzhugh and Junko Habu, pp. 53–72. Kluwer-Plenum, New York.

Hayden, Brian. 1994. Competition, Labor, and Complex Hunter-Gatherers. In *Key Issues in Hunter-Gatherer Research,* edited by Ernest S. Burch and Linda J. Ellana, pp. 223–240. Berg, Oxford.

Hudson, Mark. 2022. *Bronze Age Maritime and Warrior Dynamics in Island East Asia.* Cambridge University Press, Cambridge.

Ikeya, Nobuyuki. 2015. Maritime Transport of Obsidian in Japan during the Upper Paleolithic. In *Emergence and Diversity of Modern Human Behavior in Paleolithic Asia,* edited by Yousuke Kaifu, Masami Izuho, Ted Goebel, Hiroyuki Sato, and Akira Ono, pp. 362–375. Texas A&M University Press, College Station.

Irwin, Geoffrey, Simon Bickler, and Philip Quirke. 1990. Voyaging by Canoe and Computer: Experiments in the Settlement of the Pacific Ocean. *Antiquity* 64(242): 34–50.

Irwin, Geoffrey, Dilys Johns, Richard Flay, Filippo Munaro, Yun Sung, and Tim Mackrell. 2017. A Review of Archaeological Māori Canoes (Waka) Reveals Changes in Sailing Technology and Maritime Communications in Aotearoa / New Zealand, AD 1300–1800. *Journal of Pacific Archaeology* 8: 31–43.

Johnson, A. W., and T. K. Earle. 1987. *The Evolution of Human Societies: From Foraging Group to Agrarian State*. Stanford University Press, Stanford, California.

Kahn, Jennifer G., Abigail Buffington, Claudia Escue, and Stefani A. Crabtree. 2022. Social and Ecological Factors Affect Long-Term Resilience of Voyaging Canoes in Pre-Contact Eastern Polynesia: A Multiproxy Approach from the ArchaeoEcology Project. *Frontiers in Ecology and Evolution* 9: 750351. https://doi.org/10.3389/fevo.2021.750351.

Kelly, Robert L. 2013. *The Lifeways of Hunter-Gatherers: The Foraging Spectrum*. Cambridge University Press, Cambridge.

Kintigh, Keith W., Jeffrey H. Altschul, Mary C. Beaudry, Robert D. Drennan, Ann P. Kinzig, Timothy A. Kohler, W. Frederick Limp, Herbert D. G. Maschner, William K. Michener, and Timothy R. Pauketat. 2014. Grand Challenges for Archaeology. *American Antiquity* 79(1): 5–24.

Kirch, Patrick Vinton. 2010. *How Chiefs Became Kings: Divine Kingship and the Rise of Archaic States in Ancient Hawai'i*. University of California Press, Berkeley.

Lanting, Jan N. 2000. Dates for Origin and Diffusion of the European Logboat. *Palaeohistoria* 39/40(1997/1998): 627–650.

Lee, Sang-Mog, and Daniel Robineau. 2004. Les cétacés des gravures rupestres néolithiques de Bangu-dae (Corée du Sud) et les débuts de la chasse à la baleine dans le Pacifique nord-ouest. *L'anthropologie* 108(1): 137–151.

Lesnek, Alia J., Jason P. Briner, Charlotte Lindqvist, James F. Baichtal, and Timothy H. Heaton. 2018. Deglaciation of the Pacific Coastal Corridor Directly Preceded the Human Colonization of the Americas. *Science Advances* 4(5): eaar5040. https://doi.org/10.1126/sciadv.aar5040.

Ling, Johan, Timothy Earle, and Kristian Kristiansen. 2018. Maritime Mode of Production: Raiding and Trading in Seafaring Chiefdoms. *Current Anthropology* 59(5): 469–662. https://doi.org/10.1086/699613.

Ling, Johan, Mikael Fauvelle, Knut Ivar Austvoll, Boel Bengtsson, Linn Nordvall, and Christian Horn. 2024. Where Are the Missing Boatyards? Steaming Pits as Boat Building Sites in the Nordic Bronze Age. *Praehistorische Zeitschrift* 99(2). https://doi.org/10.1515/pz-2024-2005.

Lira, Nicolás, Valentina Figueroa, and Romina Braicovich. 2015. Informe sobre los restos de la dalca del Museo Etnográfico de Achao, Chiloé. *Magallania* 43(1): 309–320.

Marean, Curtis W., Miryam Bar-Matthews, Jocelyn Bernatchez, Erich Fisher, Paul Goldberg, Andy I. R. Herries, Zenobia Jacobs, Antonieta Jerardino, Panagiotis Karkanas, and Tom Minichillo. 2007. Early Human Use of Marine Resources and Pigment in South Africa during the Middle Pleistocene. *Nature* 449(7164): 905–908.

Montenegro, Álvaro, Richard T. Callaghan, and Scott M. Fitzpatrick. 2014. From West to East: Environmental Influences on the Rate and Pathways of Polynesian Colonization. *The Holocene* 24(2): 242–256.

Montenegro, Álvaro, Reneé Hetherington, Michael Eby, and Andrew J. Weaver. 2006. Modelling Pre-Historic Transoceanic Crossings into the Americas. *Quaternary Science Reviews* 25(11–12): 1323–1338.

Moseley, Michael Edward. 1975. The Maritime Foundations of Andean Civilization. Cummings, Menlo Park, California.

Ohtsuka, K. 1999. Itaomachip: Reviving a Boat Building and Trading Tradition. In *Ainu: Spirit of a Northern People,* pp. 374–376. Smithsonian Institution, Washington, D.C.

Osborn, Alan. 1977. Strandloopers, Mermaids, and Other Fairy Tales: Ecological Determinants of Marine Resource Utilization—The Peruvian Case. In *For Theory Building in Archaeology,* edited by L. R. Binford, pp. 157–205. Academic Press, New York.

Ruhl, Donna L., and Barbara A. Purdy. 2005. One Hundred-One Canoes on the Shore—3–5,000 Year Old Canoes from Newnans Lake, Florida. *Journal of Wetland Archaeology* 5(1): 111–127.

Sampson, Christina Perry. 2023. *Fisher-Hunter-Gatherer Complexity in North America.* University Press of Florida, Gainesville.

Sassaman, Kenneth E. 2004. Complex Hunter-Gatherers in Evolution and History: A North American Perspective. *Journal of Archaeological Research* 12(3): 227–280.

Scott, James C. 2017. *Against the Grain: A Deep History of the Earliest States.* Yale University Press, New Haven, Connecticut.

Spriggs, Matthew. 2011. Archaeology and the Austronesian Expansion: Where Are We Now? *Antiquity* 85(328): 510–528.

Strasser, Thomas F., Curtis Runnels, Karl Wegmann, Eleni Panagopoulou, Floyd Mccoy, Chad Digregorio, Panagiotis Karkanas, and Nick Thompson. 2011. Dating Palaeolithic Sites in Southwestern Crete, Greece. *Journal of Quaternary Science* 26(5): 553–560.

Vander Linden, Marc. 2016. Population History in Third-Millennium-BC Europe: Assessing the Contribution of Genetics. *World Archaeology* 48(5): 714–728.

Wengrow, David. 2018. The Origins of Civic Life—A Global Perspective. *Origini: Prehistory and Protohistory of Ancient Civilizations* XLII(2018–2): 25–44.

Westerdahl, Christer. 1985. Sewn Boats of the North: A Preliminary Catalogue with Introductory Comments. Part 2. *International Journal of Nautical Archaeology* 14(2): 119–142.

Westerdahl, Christer. 1992. The Maritime Cultural Landscape. *International Journal of Nautical Archaeology* 21(1): 5–14. https://doi.org/10.1111/j.1095-9270.1992.tb00336.x.

Widmer, Randolph J. 1988. *The Evolution of Calusa: A Nonagricultural Chiefdom of the Southwest Florida Coast.* University of Alabama Press, Tuscaloosa.

Wilmshurst, Janet M., Terry L. Hunt, Carl P. Lipo, and Atholl J. Anderson. 2011. High-Precision Radiocarbon Dating Shows Recent and Rapid Initial Human Colonization of East Polynesia. *Proceedings of the National Academy of Sciences* 108(5): 1815–1820.

Wright, Edward. 2014. *The Ferriby Boats: Seacraft of the Bronze Age.* Routledge, London.

Zilhão, J., D. E. Angelucci, M. Araújo Igreja, L. J. Arnold, E. Badal, P. Callapez, J. L. Cardoso, et al. 2020. Last Interglacial Iberian Neandertals as Fisher-Hunter-Gatherers. *Science* 367(6485): eaaz7943. https://doi.org/10.1126/science.aaz7943.

2

Navigating Paradigms

Seafaring, Settlement Patterns, and Social Interaction in Southernmost South America

Alberto García-Piquer

Seafaring has been an essential part of the daily life of the human communities that populated the seawaters of Fuego-Patagonia for at least the last 6,500 years. Archaeological evidence is indirect but solid, consisting of middle and late Holocene archaeological sites in island settings and a relatively homogeneous archaeological record through the ensuing millennia that is characterized by maritime-oriented subsistence strategies (Orquera and Piana 2009; San Román et al. 2016). Post-contact evidence for the importance of canoes in the life of Kawésqar and Yagán communities is invariably found in classic ethnographies (Emperaire 1963; Gusinde 1991), ethnohistorical sources (see Aguilera 2023; Aguilera et al., this volume: Chapter 6), oral memory (Aguilera and Tonko 2020), and the linguistic record (e.g., the recent work of Eidshaug et al. 2024).

Canoes were a keystone of the way that Fuego-Patagonian societies historically negotiated daily life with the environment and other human groups, shaping their maritime worlds. From an archaeological perspective, these multiple human–boat–environment negotiations can be studied at three different spatial scales. A first scale of relations has been noted very eloquently by Bjerck (2016), who suggested in his study of the Beagle Channel that the restrictions in transport capacity imposed by the vernacular watercraft, the bark canoe, limited the size of the social unit. This led to a long-term pattern of settlement homogeneity that can be archaeologically observed. The archaeological investigation of "site location analysis" is a way of approaching human–boat–environment relationships at a broader, although still local, scale. Different works have emphasized that site location decisions were strongly influenced by the demands of the vehicle and the configuration of the coast. In the western Patagonian archipelagos, the location of archaeological sites is strongly determined, at the local level, by their orientation in relation to the prevailing westerly winds and the availability of beaches with gentle slopes (Curry 1991; García-Piquer et al. 2025; Legoupil 2000). However, beyond this clear pattern, there are also exceptions, perhaps related to the seasonal variation of winds or to portions of the coastline that presented specific incentives. Moreover, in other parts of Fuego-Patagonia, like in the Beagle Channel, wind protection or beach slope do not appear to have been significant factors (Barceló et al. 2002). This could also be the case for the outer Pacific coast of western Patagonia (Laming-Emperaire 1972).

The third level of human–boat–environment relations has to do with the role played by landforms, currents, tides, and winds in shaping the maritime routes followed by past seafarers at a regional scale (Figure 2.1). The interplay between these elements and the technical performance and capacity of the canoes, as well as the "collective" skill and knowledge of the seafarer, is of obvious interest as it could have influenced in a very determining way settlement patterns and social interaction networks in the past. Therefore, this subject has been partially addressed by a wide range of authors and scholars since the very beginning of the archaeology and anthropology fields in Fuego-Patagonia. Drawing from the archaeological sites documented by the anthropologist J. Emperaire during his travels with the Kawésqar of Puerto Edén (Emperaire 1963) and his archaeological surveys in 1951–1952, Laming-Emperaire (1972) suggested that archaeological sites of the western Patagonian archipelagos are not randomly distributed but focused on delimited areas. Among other things, Laming-Emperaire drew attention to a "middle strip" almost without archaeological sites (see map on Figure 2.2). This "empty middle area" is centered on Nelson Strait, a large ocean entry that penetrates eastward into the archipelagos.

On the Pacific coast of the western Patagonian archipelagos, between the Gulf of Penas and the Strait of Magellan, the permanent ocean currents range on average from 10 to 25 miles per day, roughly 0.2 to 0.5 m/s, although they can have much higher intensities with strong continuous winds (Benavente 1988; Figure 2.2). Tide currents are very complex and are strongly modified by atmospheric and topographic factors, with the wind being the most influential one (Benavente 1988). The flood current runs from the ocean to the mainland (west to east), and along the longitudinal channels toward the south. Conversely, the ebb current always flows toward the ocean or toward the north. The huge volume of seawater that enters through the ocean opening of the Nelson Strait divides into four main branches, determining the tidal pattern (Figure 2.2). Sea currents in the area are relatively weak to moderate, with values that range between 0.5 and 2 knots (0.25–1 m/s). The main exception in the region would be the Kirke Narrows, which can reach velocities of up to 10 knots, or 5 m/s

Figure 2.1. A view from the boat of the complex seascapes and sometimes dangerous waters of the western Patagonian channels (the White Narrows). Photo by A. García-Piquer.

(Benavente 1988). On the contrary, the prevailing westerlies enter through the open-sea area of the Nelson Strait, which presents very high wind intensities. The mean wind speed in the strait is 9.5–10 m/s, which, according to the Beaufort Wind Scale, implies moderate waves of 1.5–2.5 m taking a more pronounced long form and many whitecaps (Barua 2019).

Unfortunately, archaeological research in the western Patagonian archipelagos is still far from offering a complete picture. Beyond pioneering work (Bird 1988; Laming-Emperaire 1972), archaeology in the western Patagonian archipelagos has focused on relatively accessible areas, like the inland seas (García-Piquer et al. 2022; Legoupil 2000; Legoupil et al. 2003; San Román et al. 2016). In the western channels, research has been almost limited to the archaeological survey of small areas around Puerto Edén (Curry 1991) or the archipelago of Madre de Dios (Legoupil and Sellier 2004) and to subsequent expeditions organized by the French Federation of Speleology. In addition to the unevenness of the archaeological record, there is also a significant bias in ethnographic observations as almost all encounters with European and later Chilean sailors and explorers were brief and happened along the maritime traffic routes followed by modern vessels (Aguilera 2023). As stated by Laming-Emperaire, "The route of the ships that go along the channels from

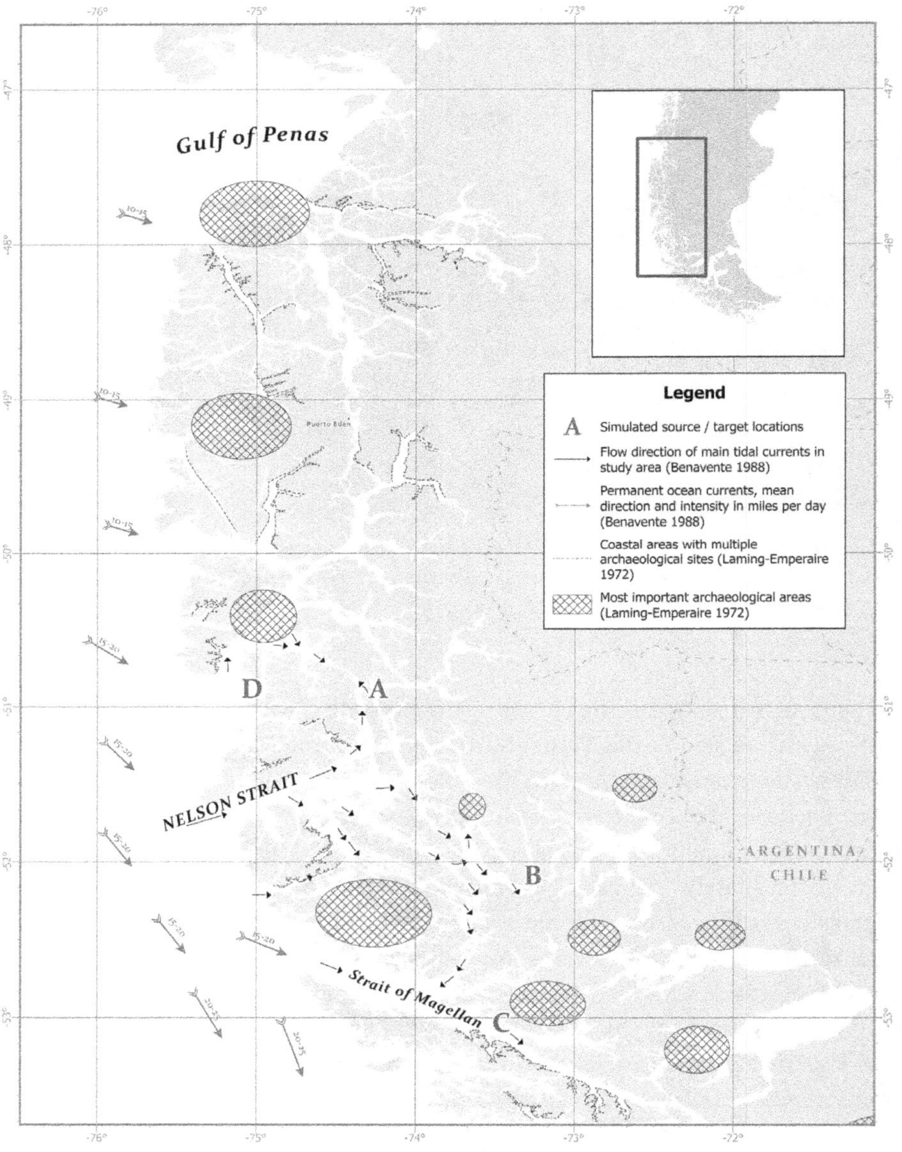

Gulf of Penas
Puerto Eden
D
A
B
C
NELSON STRAIT
Strait of Magellan
ARGENTINA
CHILE
Legend
A Simulated source / target locations
Flow direction of main tidal currents in study area (Benavente 1988)
Permanent ocean currents, mean direction and intensity in miles per day (Benavente 1988)
Coastal areas with multiple archaeological sites (Laming-Emperaire 1972)
Most important archaeological areas (Laming-Emperaire 1972)

Figure 2.2. Regional view of the western Patagonian archipelagos, with indications of sea currents from Benavente (1988), main archaeological areas from Laming-Emperaire (1972), and the sources/targets of simulated routes (letters A–D).

Puerto Montt to Punta Arenas don't correspond to the preferred habitat of the Kawésqar" (1972: 89). This spatial mismatch continues to create a biased record with significant consequences for understanding the social strategies that were implemented by seafarer peoples at the local and regional scale.

The development and application of computational methods to explore maritime mobility and routes across the complex seascapes of Fuego-Patagonian archipelagos can contribute to testing region-wide hypotheses about settlement patterns and social interaction, developing theory, and orienting research toward new areas of archaeological and anthropological interest. Prehistoric seafaring has been a fertile ground for simulations during the last decades. Many works have evaluated maritime routes with a wide range of computational methods, from calculating trajectories based on currents and winds (Fitzpatrick and Callaghan 2013; Irwin et al. 1990; Montenegro et al. 2006) to GIS-based least-cost path methods (Gustas and Supernant 2019; Perttola 2022; Slayton 2018) and, more recently, agent-based models (Hölzchen et al. 2021; Smith 2020). Frequently focused on early migrations and the colonization of new continents, these simulations have been applied in various parts of the world, but none have targeted the extreme south of South America.

This chapter evaluates the navigability of Nelson Strait using an agent-based modeling (ABM) approach to generate the most probable maritime routes that would have been traveled under specific conditions in the Pacific-central area of the western Patagonian archipelagos (Figure 2.2). In the next section, the design and main mechanics of the Hunter-Gatherer Seafaring (HUGASEA) model are presented. The results of the application of the ABM to the study region are presented and discussed in the main part of the chapter. In addition to testing the "empty middle area" hypothesis, the case study presented below provides the opportunity to discuss the importance of integrating notions of risk, planning, and amphibious mobility into the analysis of seafaring strategies. Furthermore, this chapter discusses the importance of integrating seafaring into the archaeological models of the western Patagonian archipelagos in order to understand how small-scale, highly mobile societies negotiated and built their watery worlds in the past.

The HUGASEA Model

Agent-based modeling is a complex systems approach to understanding the behavior of autonomous entities (agents) that live and interact in an artificial environment. In recent years, ABM has become a vital research approach in archaeology and anthropology (Rogers and Cegielski 2017). The computational model presented here is implemented in NetLogo 6.3.0 (Wilensky 1999).

NetLogo is a popular open-source, multiagent, programmable modeling environment that has been applied widely to simulate archaeological problems, including the study of social aggregation events or cultural diversity among Patagonian hunter-fisher-gatherers (Barceló et al. 2014; Zurro et al. 2019). Recent efforts to couple geographic information systems (GIS) with ABM have shown the benefits of joining the precision and spatial data of the former with the explicit representation of time and individual agent decision-making in ABM (Davies et al. 2019). When integrated with GIS, ABM has been used to simulate pedestrian routes over rugged terrains (Gravel-Miguel and Wren 2018) or the crossing of different sea straits (Hölzchen et al. 2021).

The HUGASEA model is an ABM that models the movement of agents paddling a canoe across a simulated environment (Figure 2.3). The objective of the model is to identify the most recurrent maritime routes between one or multiple departure locations and destinations, evaluating the degree of connectivity between given locations in terms of navigation time and risk. In the model, one or multiple agents try to reach their given destination by canoe considering the coastline configuration, the canoe velocity in calm waters, and the resistance offered by sea currents and winds. The main variables used in the model are summarized in Table 2.1. Maritime conditions are dynamic, changing every six-hour period (represented by a "tick" in the NetLogo environment). Thus, a simulated day is represented by a discrete sequence of four "ticks" (e.g., 0h, 6h, 12h, 18h). Each "tick" corresponds to annual average conditions of currents and winds. In future versions of the model, alternative maritime scenarios like rough weather or extremely calmed waters will be implemented.

The simulated world is a 1×1 km grid of "patches" (i.e., agents representing the environment). There are two main types of patches: land and water patches. Raster data containing the magnitude and direction of the velocity of currents and winds was imported independently into the model using the NetLogo GIS extension. Wind data was calculated from the Global Ocean Hourly Sea Surface Wind Model (EU Copernicus Marine Service Information; https://doi .org/10.48670/moi-00305) with 0.125 degrees spatial resolution. Currents data is derived from MOSA, an oceanographic and atmospheric forecast system developed by the CHONOS IFOP Institute (available at chonos.ifop.cl/mosa). The data has a spatial resolution that ranges from 1 to 3 square kilometers. All currents and winds data were downscaled to 1 km spatial resolution overlapping rasters. Patches without currents data were considered land patches and, thus, an obstacle to movement. The only exceptions were the patches corresponding to the two short portage routes known historically in the study area that were manually defined as transitable patches and given an arbitrary movement cost.

It is worth noting that, in this case study, the model uses present-day envi-

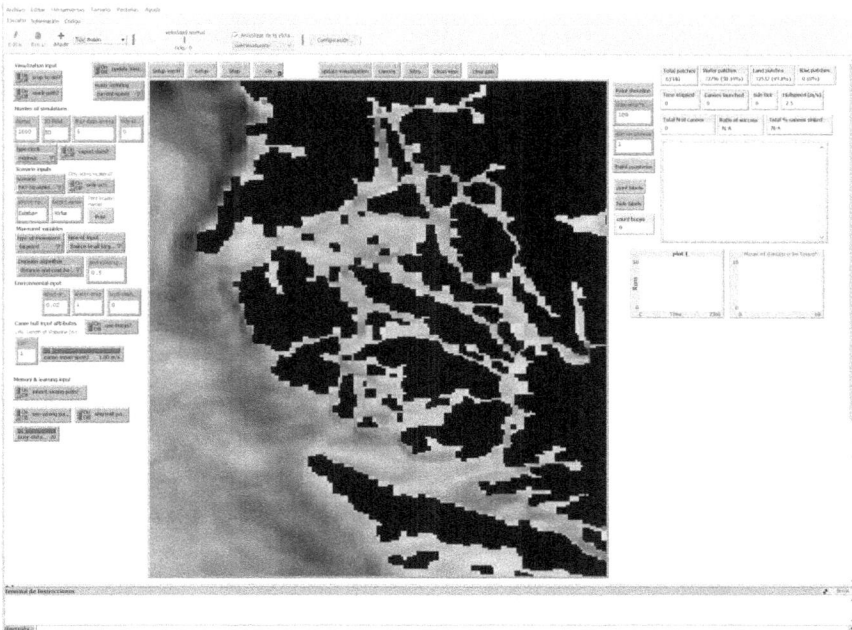

Figure 2.3. The interface of the NetLogo model with the simulated study area in the center.

ronmental conditions as input. While we can assume similar conditions for the post-contact era and, based on the location and chronology of archaeological sites, even for the last two thousand years (see, e.g., García-Piquer et al. 2025; Legoupil 2000), the feasibility of the simulated routes cannot be uncritically extrapolated to the distant past. I return to this question in the discussion.

Agents moving by canoe (referred to simply as "agents" hereafter) have an individual, short memory that allows them to remember patches visited during the same navigation, avoiding loops and wrong patches (dead ends). When an agent reaches its destination, it finishes its run and another agent is created at the source, a process repeated until the number of simulations designated by the user is reached. Canoe-agents also have a collective memory that is "stored" during the simulation in an agent-set called "buoys" and is the mechanism used by the agents to cumulatively learn and find the shortest or safest routes (depending on the chosen procedure). The mechanism works as follows: agents start the simulation acting as "wanderers"; they have a Euclidean notion of distance regarding their destination and the capacity of evaluating eight neighboring patches. After at least 10 successful runs, canoe-agents become "explorers," trying to find the shortest routes. During each navigation, "explorers" map the seascape, dropping new agent-buoys or updating existing ones. They also record information about the accumulated time or risk from the source and, if the navigation is successful, until the target. In a third stage, controlled by the variable buoy-threshold (Table 2.1), agents act as "seafarers," choosing buoys as sub-targets toward the destination, "reading" the information that has been previously stored in the buoys, and evaluating the local surroundings.

Finally, the time that a canoe-agent can paddle nonstop was adjusted with the variable max-hours-at-sea to eight hours. This value is based on the ethnographic observations of Gusinde (1991), who comments that Indigenous seafarers calculated between six and eight hours of daily navigation for long travels. A limit of eight hours for continuous paddling during long travels has also been observed empirically in the Caribbean seas (Slayton 2018). In the simulation, after eight hours of paddling, a canoe-agent must find a place to camp and rest, randomly between six and 24 hours.

The model is programmed to find either the 10 shortest paths (i.e., duration) or the 10 safest paths (i.e., risk) between locations. The simulation outcomes are exported in different formats depending on their nature. Data about agent-sets is exported using the GIS extension to shapefiles (canoes, buoys) or rasters (patches). Coordinates and associated information for each patch for the routes followed by the canoe-agents are exported as a plain text file and imported to ESRI ArcGIS Pro 3.2. Data vector conversion, visualization, and analysis are conducted in the GIS environment. All simulation input and output data are projected using WGS 84 datum.

Simulating Maritime Routes: From Least-Cost Paths to Risk-Minimizing Strategies

The results of the simulation allow us to advance some tentative conclusions that will need to be reevaluated as the agent-based model presented here is further developed and calibrated. Above all, results suggest important differences between the maritime routes selected by the agents based on two factors: the decision algorithm used and the general direction of the movement (Figure 2.4).

"Shortest paths" only consider the total duration of the route, in other words, the time that an agent-canoe would need to reach the target when following a particular route. This is computed with a score that sums the local conditions of the patches against or in favor of the canoe based on a weighted surface of cost that changes four times a day. The only limitation is given by the attributes

Table 2.1. Description of the Main Variables and Inputs Used by the HUGASEA Model for the Simulations Presented in This Chapter

Variable	Description	Input
GLOBAL VARIABLES		
Nsimul	Number of simulations for each route.	1,000
source-name	The name of the source location.	See Figure 2.1
target-name	The name of the destination.	See Figure 2.1
max-days-at-sea	Limit of total exploration time after which canoe-agents are considered "lost" and die.	Up to 50 days (automatically adjusted based on first runs)
max-hours-at-sea	Limits the number of hours paddling per day.	8 hours
tide-start	Starting moment of the route.	Random
motivation	Decision algorithm defining "optimal" route.	"Shortest path," "Safest path"
PATCH-AGENT VARIABLES		
wind-drag	The resistance generated by the winds based on the data input.	Corrected by a 0.1 factor
water-drag	The resistance generated by the water currents based on the data input.	Factor of 1 (unmodified)
portaging-speed	Velocity when portaging (fixed value).	0.5 m/s
limit-offshore	Defines open sea waters.	3 km from the shoreline
CANOE-AGENT VARIABLES		
canoe-mean-speed	Average velocity of the canoe in calm water.	1 m/s, 1.5 m/s
LWL	Length of the water level.	4 m, 7 m
BUOY-AGENT VARIABLES		
Buoy-threshold	Sets in what moment of the simulation canoe-agents begin to use learned paths.	400 simulations
Buoy-distance	The "visibility" distance range of the canoe-agents when searching buoys.	10 to 20 km
Buoy-angle	The "visibility" cone of the canoe-agents when searching buoys.	270°

of the canoe, forcing the agents to move with a physically plausible velocity (see variable LWL, Table 2.1). On the other hand, "Safest routes" considers the total risk assumed by the seafarers when following a certain route. The total risk of a route is largely a function of the time, that is, of the local resistance of currents and winds. In this sense, the longer the route, the greater the risk (Figure 2.5). The main difference in relation to the former mode—and in general to

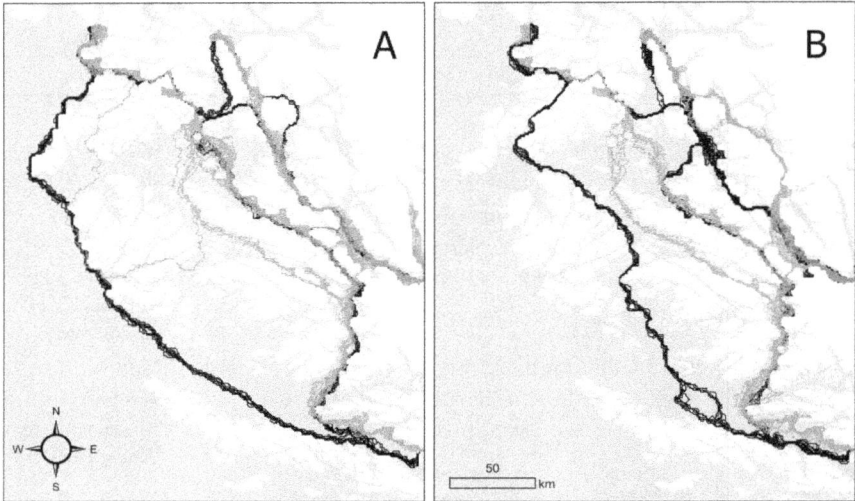

Figure 2.4. A summary view of all the maritime routes between locations: (*A*) Shortest routes; (*B*) Safest routes. Black lines correspond to southward routes; gray to north-ward routes.

pathfinding algorithms (e.g., A* algorithm) or surface-cost approaches—is that there is a perception of risk. This perception is implemented in two ways. First, there is a "threshold-value" that defines what is a dangerous situation. Drag resistance that is strong enough to deviate the canoe from its heading (assumed between -20° and 20°) is considered dangerous. Thus, slight differences between relatively safe local conditions have less weight in the simulation. Second, there is a global score that considers the number of patches (units of 1 km) that present harsh or dangerous conditions, such as open/unprotected waters and big waves. When the option is activated, the agent will tend to choose sections of the routes with less "dangerous" patches during the simulation.

A comparison between the best 10 shortest (fastest) simulated routes and the 10 best "safest" routes is shown in Figure 2.4. Although some routes or sections of routes are highly conditioned by the topography of the channels regardless of the search algorithm used, routes change dramatically in the Pacific western coast and in the eastern inner channels. In addition, maritime routes followed are highly dependent on the direction of the movement because of the broad pattern created by the west-southwest direction of winds in the region. In that sense, while the intensity of winds affects the route chosen locally, forcing agents to avoid unprotected waters, the direction of winds (and tide currents)

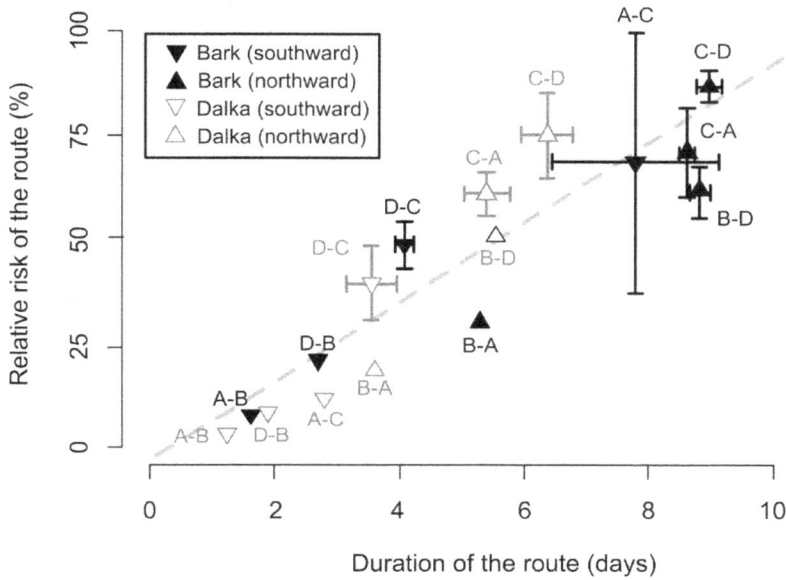

Figure 2.5. A summary graphic of the duration and risk of the simulated (safest) routes. Labels correspond to the locations indicated in Figure 2.1.

impacts the routes across the Nelson Strait. In general, moving northward is significantly slower and faces more resistance than moving southward (Figure 2.5). As I discuss further on, the two portage routes implemented in the simulation, particularly the portage of Staines Peninsula, were key in this network of routes.

Finally, there are important differences between the results depending on the type of watercraft tested. Based on the ethnohistorical record and the preserved canoes in museum collections, plank canoes were longer, increasing the length of the water line and the number of paddlers (see Aguilera 2023; Aguilera et al., this volume: Chapter 6, particularly Figure 6.6). Based on these assumptions, an alternative scenario was simulated considering different attributes of the vessel (Table 2.1). Results indicate that a slightly faster canoe can dramatically impact the routes in terms of the total time needed but also the risk that represents, in some cases offering new routes (Figure 2.5). In the case of medium to long routes, differences can be on the order of two days. Therefore, historical changes in watercraft technology could have altered previous conditions, allowing for covering longer distances per day and, therefore, altering the spatial relations of the traditional seascape. The archaeological implications of these changes in transport technology are discussed below.

Beyond Maritime–Terrestrial Dichotomy: The Practical and Social Importance of Portage Routes

Ethnohistorical sources suggest that portage routes, known locally as *"pasos de indios,"* were a keystone of Fuego-Patagonia mobility strategies, shortening travel distances and providing safer routes in terms of seafaring conditions (Prieto et al. 2000). They also enabled movement across a highly complex mosaic of microenvironments and changing weather patterns (Curry 1991). Although the use of long land routes is well attested, like the ones across the Isthmus of Ofqui (Byron 1768) or Lapataia valley (Kent 1924), most of the portages identified in ethnographic sources in the western Patagonian archipelagos are short routes across a narrow isthmus (see Aguilera 2023; Maximiano Castillejo 2017).

Two portage areas have been identified in the study region. The first one is a 200-meter-wide isthmus connecting the bays of Isthmus and Oración, allowing the portage of canoes between Smyth Channel and Union Sound. The ethnohistorical record contains multiple encounters between Indigenous and white travelers that used this well-known area to portage their boats (Serrano Montaner 1891; for a summary, see Aguilera 2023). The other one is a 1-km-wide isthmus between the Staines Peninsula and the glaciers of the Southern Patagonian Ice Field. As this area lies outside the main waterways used by modern sailors, little is known about it. It has been proposed as a possible portage route based on its geographical characteristics and strategic location (Prieto et al. 2000). The only reference known by the author of this chapter was generated within the framework of a geological survey conducted in the area during the 1970s: "Long after, during a geological expedition in the area, we discovered that they [the Indigenous people] cross the Staines peninsula overland, going up from the Yussef Sound and to the north, getting out in front of the Owen Island" (Stäger 2016: 52).

The results of the simulation indicate that both portages played a strategic role in the routes across the Nelson Strait, providing safer passages (Figure 2.6). Based on the results, the Isthmus Bay portage would have provided the opportunity to switch to less exposed coasts and take advantage of the north–south clash of currents around Collingwood Strait (Benavente 1988) when going to or coming from the Strait of Magellan. The portage of Staines, on the other hand, would have represented a safer alternative to the route of Smyth

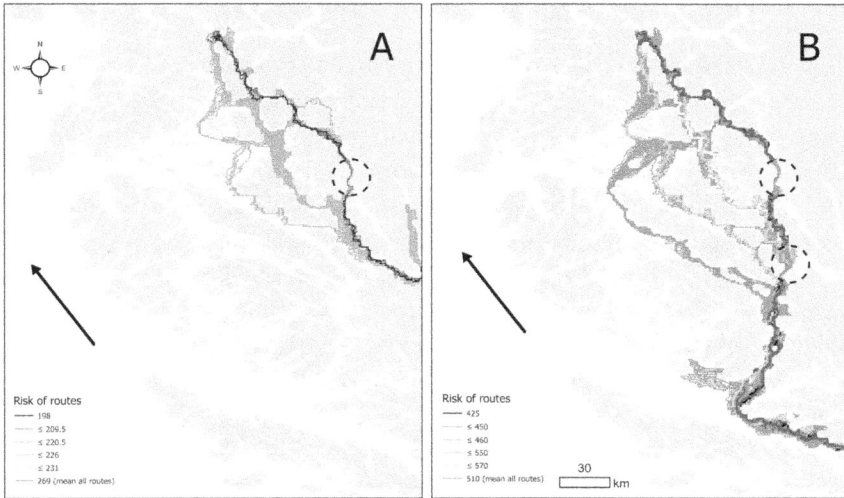

Figure 2.6. Simulated risk (indicated by color) of the maritime routes connecting the inland sea of Última Esperanza (*A*) and the Magellan Strait (*B*) with northwestern channels. The direction of the routes is northward (upwind).

Channel when traveling south to north (Figure 2.6). Significantly, these results are in accordance with the observation made by Stäger (see above).

It is difficult to verify these results archaeologically, not only because archaeological work in the area is very limited but also because the study of portage routes is challenging for archaeology. Resting, unloading/loading, or repairing the watercraft are activities that can be expected in portages (Prieto et al. 2000). Indigenous huts or campsites have been historically observed along some long routes (Kent 1924), and archaeological surveys have documented shell middens at the beginning of multiple-day routes (Prieto et al. 2000). However, one may expect that the probabilities of finding archaeological evidence of short portages are less than in longer ones. Despite ethnohistorical evidence, a short visit to the area of Isthmus Bay did not find any archaeological evidence (Legoupil et al. 2003). The characteristics that made this isthmus a good portage pass in recent times, particularly its low and flat elevation profile (Maximiano Castillejo 2017), may have contributed to poor preservation of archaeological sites close to the shoreline, which is considerably impacted by marine erosion (Legoupil et al. 2003). On the other hand, the portage of Staines is highlighted as an archaeologically relevant location within the "middle empty area" based on the map of Laming-Emperaire (1972; see Figure 2.2). As I discuss in the next section, this could suggest that it is a place with multiple "passage" sites.

Archaeological Implications of Navigating along Traditional Routes

Drawing from the archaeological sites documented by Emperaire during his hunting expeditions with the Kawésqar, Laming-Emperaire (1972) presented a first archaeological model of the archipelagos of western Patagonia. Despite being a very coarse-grained model, it provides information about the western Patagonian archipelagos, still a very unknown region for archaeology. In this work, Laming-Emperaire distinguishes between "areas with permanent settlement" and "transit areas," remarking that areas with a high density of archaeological sites correspond to passage areas with multiple and dispersed small, short occupations. In her own words: "Passage sites corresponded to a settlement dispersion of very small groups during very short periods of time. Probably, the circles in the map [see Figure 2.2, this chapter] mainly indicate multiple sites of passage." Furthermore, Laming-Emperaire adds that zones without archaeological evidence could be explained either by the environmental conditions or by their closeness to more permanent sites: "The Alakaluf were able to travel by canoe very long distances in just one day, so they were not at all interested in building a new dwelling close to the permanent sites" (Laming-Emperaire 1972: 90; translations are mine).

During the last decades, archaeological investigations have supported Laming-Emperaire's statement that archaeological settlement patterns of the western Patagonian archipelagos are not randomly distributed but focused on delimited areas (Curry 1991; García-Piquer et al. 2022; Legoupil 2000). However, a common underlying assumption in settlement models is that a higher density of sites equates to the greater centrality or importance of the area within a regional settlement pattern. On the contrary, the definitions of "transit areas" and "more permanent areas" by Laming-Emperaire (1972) suggest the inverse correlation. Areas rich in archaeological sites would be "transit areas," while the preferred settlement areas of the Kawésqar, at least in recent times, would be characterized by a low density of archaeological sites. Local and regional configuration, duration, centrality, and recurrence of the human occupation emerge as intersecting dimensions that define a settlement system composed of different settlement patterns (Figure 2.7).

For other rugged seascapes like those of the Northwest coast of North America, it has been proposed that centrality of residential sites is a fundamental characteristic of seafaring mobility and settlement systems (Ames 2002). Despite the significant differences between Northwest coast and Fuego-Patagonian social and organizational strategies, centrality also emerges as a key factor of winter residential pattern in Laming-Emperaire's observations. Being able to "travel by canoe very long distances in just one day" (Laming-Emperaire 1972:

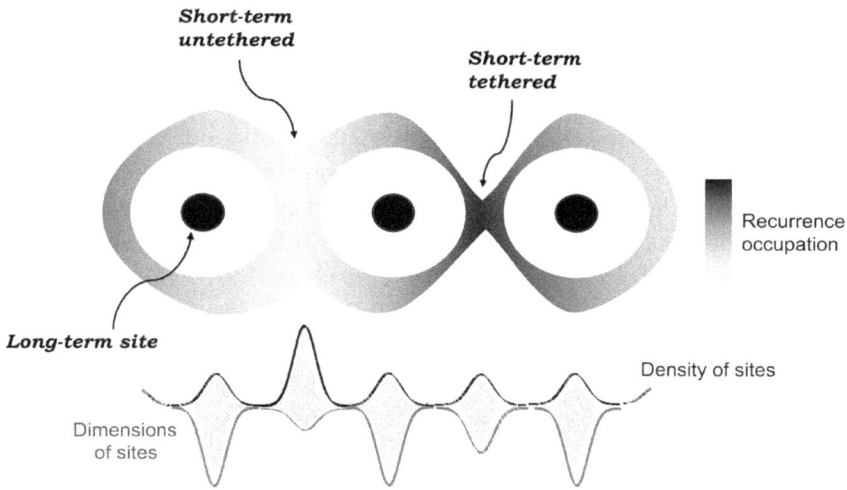

Figure 2.7. A diagram of the Kawésqar settlement system based on Laming-Emperaire (1972). The top layer represents the recurrence of occupation and the bottom layer the density and dimensions of the sites. Due to the availability of watercraft, short-term sites are not expected close to the "long-term" (more recurrent) occupations. In addition, the density of short-term sites (and thus their visibility) would depend on the constrictions created by the seascape.

90), the use of canoes would have allowed movement between otherwise distant ecological niches and connections between preferential sites, giving a pattern of "empty spaces." In turn, these mobility strategies would have required suitable navigation and docking conditions. The number of passage sites would have depended on the geomorphological characteristics of the seascape. Intensive use of the territory resulting from geographical and topographical factors would result in a lower site density (see Figure 2.6). In this system, human–boat–environment negotiations, and not the distribution of resources, define the settlement pattern.

Kawésqar language distinguishes between two large biogeographical areas in the northern sector of the western Patagonian archipelagos: *jáutok*, the territory of the inner channels projecting eastward toward the mainland, character-ized by steep coasts with narrow and rocky beaches; and *málte*, the coast near the Pacific Ocean, where beaches are generally long and sandy (Aguilera and Tonko 2013). At least during the nineteenth and twentieth centuries, northern Kawésqar took advantage of this variation, integrating the two marine worlds, *jáutok* and *málte*, into a seasonal mobility system. Indeed, the Pacific coast was visited in spring–summer to exploit the abundant colonies of sea birds and sea lions during bird breeding season and pupping season; when the weather worsened, people navigated back to the inner coasts. This mobility system generated a scattered short-term settlement pattern on the Pacific coast during spring–summer. The rest of the year people would have periodically reoccupied a few privileged residential areas located far away from each other, traveling "very considerable distances in just one day" and eventually stopping one or two nights in passage sites (Laming-Emperaire 1972: 90, translation is mine).

The customs and practices followed or remembered by the 1950s' Kawésqar community of Puerto Edén cannot be extrapolated to interpret the archaeo-logical record of western Patagonia or the whole Fuego-Patagonian archi-pelagos. Moreover, the physical geography of the landscapes and coasts that Fuego-Patagonian peoples negotiated in the distant past has been transformed drastically since the Last Ice Age. The positions of shorelines have changed dramatically since the icesheet that covered most of the region began to retire 21,000 to 15,000 years ago (Davies et al. 2020). While regional relative sea-level history since the Late Glacial Maximum is still poorly understood, we know that the rugged coastlines of southernmost South America underwent complex processes that involved the interaction of isostatic, eustatic, and tectonic forces and produced locally specific and regionally heterogeneous relative-sea-level histories. While a detailed account of relative-sea-level evolution and regional variation is beyond the scope of this chapter, it should be noted that postglacial sea-level rise reached its peak in the middle Holocene, when shorelines were located between 3 and 12 m above present sea levels, according to the few works available for the archipelagos of western Patagonia (Jaillet et al. 2008). Due to the isostatic rebound and based on the location and chronology of archaeological sites, sea levels dropped gradually until reaching levels very close to present ones around 2,000 years ago (see, e.g., García-Piquer et al. 2025; Legoupil 2000).

Kawésqar strategies suggest a level of complexity in the organizational strat-egies of Fuego-Patagonian seafarers that was generally underappreciated or omitted by non-Native observers, who only saw "maritime opportunistic wan-derers," and that can be evaluated against the archaeological record of the later Late Holocene. Northern Kawésqar unfolded both planned and opportunistic strategies along traditional routes of navigation and across charted waters, tak-ing advantage of their deep knowledge of the seascape and probably extensive networks of information (Aguilera and Tonko 2020). While opportunistic or foraging strategies would have predominated in certain seascapes and moments of the year, periodic reoccupation of distant, privileged areas would have favored more "logistical" or planned strategies.

The Nelson Strait: A Geographical Boundary or a Crossroad?

Ethnographer Martin Gusinde identified three dangerous open-sea areas that would have acted as "barriers" for Kawésqar populations, from south to north: the Pacific entrance of the Strait of Magellan, the Nelson Strait, and the Gulf of Penas (Figure 2.2). These "natural boundaries" would have separated northern, central, and southern Kawésqar partialities or groups (Gusinde 1991: 125). In addition to Laming-Emperaire's work (1972), more recent work has evaluated the environmental conditions of the western entrance of the Strait of Magellan (Pallo 2011) and the Gulf of Penas (Reyes et al. 2022), highlighting their role as geographical and cultural barriers throughout the history of Fuego-Patagonian seafarer peoples.

However, ethnohistorical sources and life stories leave no doubt that the Kawésqar were skillful seafarers who frequently navigated across dangerous waters, even at very unprotected areas of the Pacific coastline (Aguilera 2017; Aguilera and Tonko 2020). Ethnic and language analysis based on ethnohistorical sources indicates that the Gulf of Penas was more a common space than a barrier, at least in recent times (Aguilera 2023). Archaeological evidence is limited but enough to attest that the expertise and boldness of ancestral populations was not less. At least 4,000 years ago, prehistoric seafarers were living at or traveling to the archipelago of Madre de Dios to ceremonially dispose of their dead (Legoupil and Sellier 2004). Even the southernmost island of Cape Horn was visited by Yagán seafarers in recent times, perhaps in a bark canoe, despite the highly exposed location (Buma et al. 2022). The critical question, then, is not whether Indigenous populations were able to travel across the Nelson Strait but how, when, and along which routes they traveled.

From the point of view of latitudinal movement across the western Patagonian archipelagos, the Nelson Strait lies at the center of a cultural crossroads (Figure 2.8). According to oral tradition, this region was inhabited by Kawésqar groups known as Keláelkčes, literally the "inhabitants of Keláel." Keláel is referred to as the inland sea of Última Esperanza (Aguilera 2017). The geographical layout of the Nelson Strait, together with the entrance of the Magellan Strait to the south, was not a barrier to Indigenous seafarers but strongly conditioned maritime routes, affecting settlement patterns and social interaction. Indeed, the results of the simulation suggest that mobility across the area would have been funneled through very specific channels, depending on the direction of the movement. It is possible to hypothesize that the general direction of the wind would have made it easier to move southward on the outside coast, and northward in the inside channels, promoting a roughly circular pattern (Figure 2.8).

This pattern would explain the "middle empty area" noted by Laming-Emperaire (1972). Furthermore, the routes that avoid open waters pass through two of the "most important archaeological areas." Following the settlement model of Laming-Emperaire and the *jáutok-málte* definition of the seascape, these can be interpreted as dispersed (west) or concentrated (east) clusters of passage sites. The portage of Staines has a central role in this system, both geographically and socially. The role of the Isthmus portage is more complicated to evaluate. While its use by Indigenous and non-Indigenous seafarers is well attested in the ethnographic record, perhaps it was more important to the latter than the former. Traditionally, Indigenous seafarers would have alternative routes, like the routes across the inland seas (Figure 2.8). In addition, they would have been more interested in camping closer to the more protected waters of the Union Sound and the Kirke Narrows. Indeed, it is known that the best way to cross the dangerous Kirke Narrows is during slack water, with a duration of 15 min (high tide) or 30–45 min (low tide; Benavente 1988). Slack-water navigation of this passage and the White Narrows, which represent the only maritime connections between the inland sea of Última Esperanza and the western channels, could explain the clustering of sites in bays around them (García-Piquer et al. 2022).

Conclusion and Further Steps

From our human perspective of time and history, landforms are relatively immutable agents, embedding settlement, space, and social relationships. Seawaters represent connections, change, and fluidity, and they manifest as a dynamic organism that can mutate completely within hours or even minutes. Many past human societies not only learned to move across the seas but placed their energetic and spirited nature at the center of their social and economic life. Knowledge about time and space, being aware of changes, reading the sea, the trees, the sky—all are key components of maritime movement. Thus, simulating maritime routes can become a considerable challenge, sandwiched between oversimplification and an ocean of complexities.

In this chapter, I have applied the HUGASEA model, an agent-based simulation approach, to explore how geography and natural forces such as the ebb and flow of tides, winds, and currents conditioned the duration and feasibility of maritime routes across the central part of the western Patagonian archipelagos. The Nelson Strait, an area that has been described as a "middle empty area" (Laming-Emperaire 1972) or a "geographical barrier" (Gusinde 1991) to Kawésqar seafarers and their ancestors, provides an almost ideal test case scenario as it is surrounded by a maze of channels and delimited to the east by

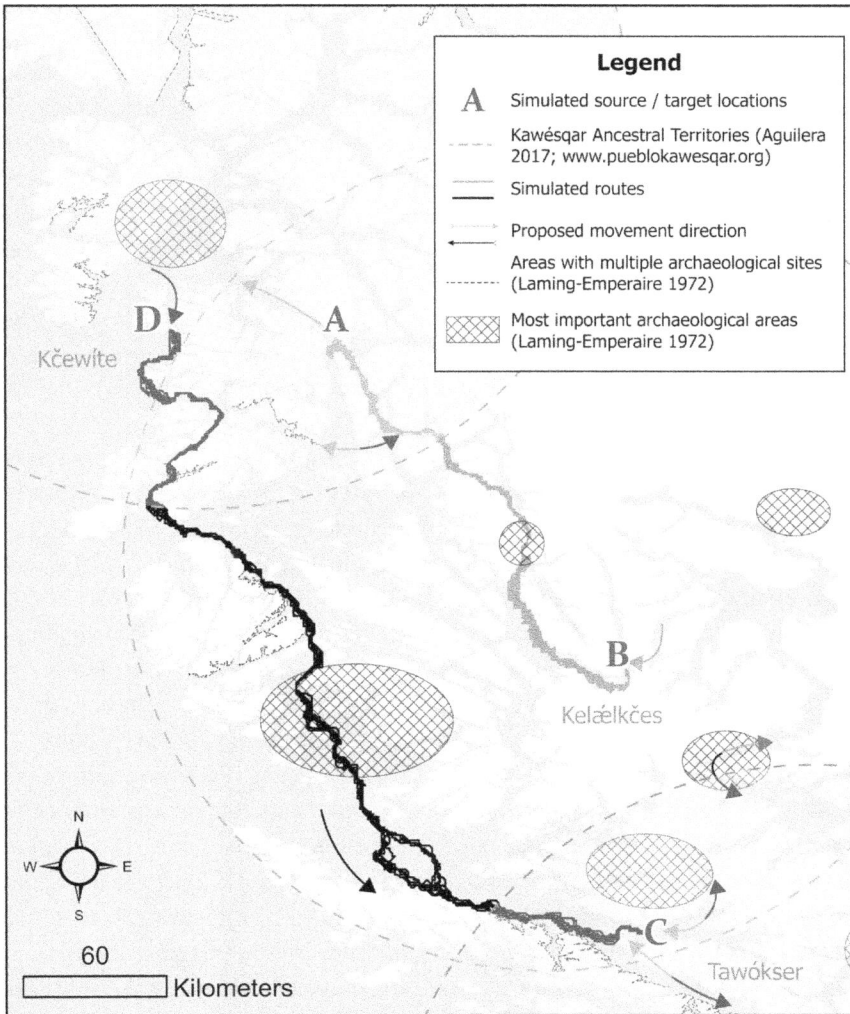

Legend

A Simulated source / target locations

---- Kawésqar Ancestral Territories (Aguilera 2017; www.pueblokawesqar.org)

──── Simulated routes

──→ Proposed movement direction

------- Areas with multiple archaeological sites (Laming-Emperaire 1972)

▦ Most important archaeological areas (Laming-Emperaire 1972)

Figure 2.8. A maritime circulation model for the region of Keláel based on the simulation results and incorporating archaeological and historical data.

the landforms of the Southern Patagonian Ice Field and the Muñoz Gamero Peninsula. The crossing of bodies of water has been a common topic in the field of maritime simulations, but the approach presented in this chapter explores some aspects that are not as frequently integrated in this type of simulation: the notion of risk, amphibious mobility, the use of different Indigenous watercrafts, and settlement patterns.

The results presented and discussed in this chapter present some limitations, both extrinsic and intrinsic to the agent-based model being used. For the former, the IFOP (Instituto de Fomento Pesquero) is still in the experimental phase with its oceanographic and atmospheric forecast system from which the currents data has been extracted, and the availability of the temporal data is not representative of a whole year. Despite this, general daily patterns were verified against maritime records (e.g., Benavente 1988), and general daily trends are similar. More critical, current data are missing for the narrowest channels between the western islands of the study area. In some but not all cases, these data have been manually corrected. Regarding intrinsic limitations, the HUGASEA model is still in development. Each route and scenario has been simulated 1,000 times and replicated several times. Results are solid, but more sensitivity analysis is needed.

Furthermore, there are multiple factors that must be considered when discussing notions of duration, cost, risk, or even the best route in navigation. Both universal and historical factors affect these notions as well as individual and collective "perceptions" (see Jarrett, this volume: Chapter 12). The way of evaluating seafaring risk proposed here must be seen, thus, as an operating simplification. Ethnographers noted that Indigenous seafarers using bark canoes normally tried to navigate the protected waters with kelp. However, there are multiple historical accounts of Indigenous navigation across open waters over short distances. And a few observations record journeys between islands separated by 17 km or even 26 km of open sea (Orquera and Piana 1999: 293). The question remains to evaluate in what conditions—where and when—open-sea navigation was not only possible but preferable. The duration of the routes across the Nelson Strait would have increased dramatically based on the direction of the movement as agents had to confront strong headwinds. These results accord with historical observations: in the Beagle Channel, and regardless of their urgency, a group of Indigenous seafarers needed four days to cover a route of 50 km due to the windy conditions (Aspinall 1888: 31, cited in Orquera and Piana 1999: 293).

Unlike the more stable dominant wind direction, tide currents affect locally and in various ways the routes chosen by the agents. In this case, time—not space—is critical. One of the key elements of the HUGASEA model is that agents learn collectively through the exploration of previous agents. Contrary to top-to-bottom approaches, like least-cost path analysis, agents start with the capacity of evaluating only their immediate surroundings, incorporating global knowledge of the marine environment through the simulation. This allows the agent to find the shortest or safest route between locations, avoiding or taking advantage of the flow/ebb currents' direction, regardless of the starting moment of travel. The scale of the time used in the simulation allows us to evaluate the general pattern in the study area. Future simulations will focus on smaller regions, both in the Strait of Nelson and in adjacent areas, to implement a coarser scale of time and a higher spatial resolution. This would provide the opportunity to evaluate, for example, how slack water or slack tide moments, short periods of time when the current associated with tides ceases, are integrated in maritime routes and settlement patterns. Results lead the way to explore other scenarios and test hypotheses that involve the crossings of bodies of water in Fuego-Patagonia, like the intriguing evidence of pre-European human activity in the Falkland Islands (Hamley et al. 2021). Indeed, this would be the only time in the Americas outside of the Caribbean in which islands would have been reached after losing sight of land.

Results also highlight the importance of integrating amphibious mobility in seafaring simulations. Indeed, portages are considered cultural products of great anthropological interest (Westerdahl 2004). Despite that, maritime simulation models have disregarded amphibious mobility mainly because they target unobstructed surfaces of water, such as oceans or large straits. On rugged and complex seascapes like the Fuego-Patagonian archipelagos, however, portaging was a cornerstone of mobility. The maritime routes simulated and presented here suggest that the two portages known in the study area were important to mobility, particularly the one of Staines Peninsula, which offered a safer alternative route across the region. Nevertheless, a glance at the geography of the study region and the whole Fuego-Patagonian archipelago suggests the existence of many other possible portages. A further challenge is to implement amphibious mobility in a way that agents can find unmapped portage routes.

The study case has allowed me to discuss the social and historical dimensions of mobility in the central part of the western Patagonian archipelagos. This is the first work that applies a maritime simulation approach to try to evaluate the most probable or recurrent routes based on certain criteria. While the use of present-day geography and environmental data limits the application of the results to the distant past, they show the importance of conducting, improving, and extending this type of approach to the Fuego-Patagonian archipelago. Oceanographic features such as the Nelson Strait did not represent a geographical and cultural barrier to Kawésqar peoples. However, these features and the set of maritime conditions acted as bottlenecks, structuring past settlement. And, at least in recent times, they were borderlands (sensu Reid 2015), spaces shared and inhabited by distinct Kawésqar peoples. Moreover, in high-mobile, boat-centered societies, where maritime travel facilitates covering considerable distances, "empty spaces" are a critical element for understanding settlement patterns and social interaction networks (Ames 2002). As I have emphasized throughout this chapter, a seascape is a social product, the outcome of the negotiations between space (geography and time), organizational strategies (mobility, settlement, subsistence strategies), and transport technologies (watercraft). New transport technologies may shorten the distances between locations, configuring new seascapes and social relationships while continuing to follow traditional navigation routes shaped through millennia. In turn, a seafaring system might be implemented across different coastal morphologies, altering its archaeological signal. Studying how all these different dimensions intersect, combine, transformed historically, and change through time is important to understand how a particular watery world in the past was built, perceived, and engaged.

Acknowledgments

I would like to thank Alfredo Prieto for sharing his thoughts about the "empty middle area" and, particularly, the passage reference about the Staines portage.

This work has been developed in the framework of the project HUGASEA, funded by the European Union's Horizon Europe research and innovation program under the Marie Sklodowska-Curie grant agreement No. 101066298. Views and opinions expressed are, however, those of the author only and do not necessarily reflect those of the European Union or the European Research Executive Agency. Neither the European Union nor the granting authority can be held responsible for them.

References Cited

Aguilera, Nelson. 2023. Aqa K'énak: Navegación Indígena en Patagonia 1520–1960. Una aproximación desde la etnohistoria y la arqueología. PhD dissertation, Departament de Prehistòria, Universitat Autònoma de Barcelona, Bellaterra.

Aguilera, Oscar E. 2017. El nombre Kawésqar, un problema no solo lingüístico. *Magallania* 45(1): 75–84.

Aguilera, Oscar, and José Tonko. 2013. *Relatos de viaje kawésqar. Nómades canoeros de la Patagonia Occidental.* Temuco: Ofqui Editores.

Aguilera, Oscar E., and José Tonko P. 2020. *Gente de los canales: Historias de vidas de los Kawésqar.* CONADI, Punta Arenas.

Ames, K. M. 2002. Going by Boat: The Forager-Collector Continuum at Sea. In *Beyond Foraging and Collecting: Evolutionary Change in Hunter-Gatherer Settlement Systems,* edited by Ben Fitzhugh and Junko Habu, pp. 19–52. Kluwer Academic/Plenum Publishers, New York.

Barceló, Juan Antón, Florencia del Castillo, Ricardo del Olmo, Laura Mameli, F. J. Miguel Quesada, David Poza, and Xavier Vilà. 2014. Social Interaction in Hunter-Gatherer Societies: Simulating the Consequences of Cooperation and Social Aggregation. *Social Science Computer Review* 32: 417–436.

Barceló, Juan Antón, Ernesto Luis Piana, and Daniel R. Martinioni. 2002. Archaeological Spatial Modelling—a Case Study from Beagle Channel (Argentina). In *Archaeological Informatics: Pushing the Envelope. CAA2001. Computer Applications and Quantitative Methods in Archaeology. Proceedings of the 29th Conference, Gotland, April 2001,* edited by Göran Burenhult, 351–360. British Archaeological Reports, Oxford.

Barua, Dilip K. 2019. Beaufort Wind Scale. In *Encyclopedia of Coastal Science,* edited by C. W. Finkl, and C. Makowski, pp. 1–3. Encyclopedia of Earth Sciences Series. Springer, Cham.

Benavente Mercado, Roberto. 1988. Corrientes en la Zona Austral de Chile. *Revista de Marina* 105(786).

Bird, Junius B. 1988. *Travels and Archaeology in South Chile.* Edited by John Hyslop. University of Iowa Press, Iowa City.

Bjerck, Hein B. 2016. Settlements and Seafaring: Reflections on the Integration of Boats and Settlements Among Marine Foragers in Early Mesolithic Norway and the Yámana of Tierra del Fuego. *Journal of Island and Coastal Archaeology* 12(2): 276–299.

Buma, Brian, Flavia Morello, Karina Rodriguez, and Alberto Serrano Fillol. 2022. Evidence for the Southernmost Pre-Industrial Human Expansion on Isla Hornos (Isla Lököshpi), Chile. *Antiquity* 96(389): 1324–1329.

Byron, John. 1768. *The Narrative of the Honourable John Byron (Commodore in a Late Expedition Round the World) containing an account of the Great Distresses Suffered by Himself and His Companions on the Coast of Patagonia, From the Year 1740, till their Arrival in England, 1746.* S. Baker and G. Leigh, London.

Curry, Patricia J. 1991. Distribución de sitios e implicaciones para la movilidad de los canoeros en el canal Messier. *Anales del Instituto de la Patagonia* 20: 145–154.

Davies, Benjamin, Iza Romanowska, Kathryn Harris, and Stefani A. Crabtree. 2019. Combining Geographic Information Systems and Agent-Based Models in Archaeology: Part 2 of 3. *Advances in Archaeological Practice* 7(2): 185–193.

Davies, Bethan J., Christopher M. Darvill, Harold Lovell, Jacob M. Bendle, Julian A. Dowdeswell, Derek Fabel, Juan-Luis García, et al. 2020. The Evolution of the Patagonian Ice Sheet from 35 Ka to the Present Day (PATICE). *Earth-Science Reviews* 204: 103152.

Eidshaug, Jo Sindre P., Hein B. Bjerck, Terje Lohndal, and Ole Risbøl. 2024. Words as Archaeological Objects: A Study of Marine Lifeways, Seascapes, and Coastal Environmental Knowledge in the Yagán–English Dictionary. *International Journal of Historical Archaeology* 28: 722–766.

Emperaire, Joseph. 1963. *Los Nómades del Mar.* Ediciones de la Universidad de Chile, Santiago de Chile.

Fitzpatrick, Scott M., and Richard T. Callaghan. 2013. Estimating Trajectories of Colonisation to the Mariana Islands, Western Pacific. *Antiquity* 87(337): 840–853.

García-Piquer, Alberto, Vanessa Navarrete, Nelson Aguilera, Robert Carracedo, Anna Franch, Christian García P., Eva Ros-Sabé, Gabriel Zegers, Alfredo Prieto, and Raquel Piqué. 2022. En el mar interior de Última Esperanza: Dinámicas de ocupación y movilidad canoera en la isla Diego Portales (Magallanes, Chile). *Latin American Antiquity* 33(4): 838–856.

García-Piquer, Alberto, Vanessa Navarrete, Raquel Piqué, and Alfredo Prieto. 2025. Moving across the Seascape-Landscape: Site Location Patterns in the Inland Sea of Última Esperanza (Magallanes Region, Chile) During the Late Holocene. In *HOMER 2021: Archéologie des peuplements littoraux et des interactions Homme/Milieu en Atlantique nord équateur.* Sidestone Press Academics, Leiden.

Gravel-Miguel, Claudine, and Colin D. Wren. 2018. Agent-Based Least-Cost Path Analysis and the Diffusion of Cantabrian Lower Magdalenian Engraved Scapulae. *Journal of Archaeological Science* 99: 1–9.

Gusinde, Martin. 1991. *Los indios de Tierra del Fuego: 3. Los Halakwulup.* Vol. 1. Centro argentino de etnología americana, Buenos Aires.

Gustas, Robert, and Kisha Supernant. 2019. Coastal Migration into the Americas and Least Cost Path Analysis. *Journal of Anthropological Archaeology* 54(October 2018): 192–206.

Hamley, Kit M., Jacquelyn L. Gill, Kathryn E. Krasinski, Dulcinea V. Groff, Brenda L. Hall, Daniel H. Sandweiss, John R. Southon, Paul Brickle, and Thomas V. Lowell. 2021. Evidence of Prehistoric Human Activity in the Falkland Islands. *Science Advances* 7(44): p.eabh3803.

Hölzchen, Ericson, Christine Hertler, Ana Mateos, Jesús Rodríguez, Jan Ole Berndt, and Ingo J. Timm. 2021. Discovering the Opposite Shore: How Did Hominins Cross Sea Straits? *PLOS One* 16(6): e0252885.

Irwin, Geoffrey, Simon Bickler, and Philip Quirke. 1990. Voyaging by Canoe and Computer: Experiments in the Settlement of the Pacific Ocean. *Antiquity* 64(242): 34–50.

Jaillet, Stéphane, Maire Richard, Bréhier Franck, Despain Joël, Lans Benjamin, Morel L., Pernette Jean-François, Ployon Estelle, and Tourte Bernard. 2008. Englacement, eustatisme et réajustements karstiques de la bordure sud de l'archipel de Madré de Dios (Patagonie, Province Última Esperanza, Chili). *Karstologia: Revue de karstologie et de spéléologie physique* 51: 1–24.

Kent, Rockwell. 1924. *Voyaging Southward from the Strait of Magellan.* Halcyon House, New York.

Laming-Emperaire, Annette. 1972. Los Sitios Arqueológicos de los Archipiélagos de Patagonia occidental. *Anales del Instituto de la Patagonia* 3(1–2): 87–96.

Legoupil, Dominique. 2000. El sistema socioeconómico de los nómades del mar de Skyring (Archipiélago de Patagonia). *Anales del Instituto de la Patagonia.* Serie Ciencias Humanas 28: 81–119.

Legoupil, Dominique, Marianne Christensen, Matthieu Langlais, Sébastien Lepetz, and Kai Salas. 2003. *Les voies de peuplement des archipels de Patagonie: Région d'Última Esperanza et île de Chiloé.* Rapport 2003. Mission Archéologique Française de Patagonie, Paris.

Legoupil, Dominique, and Y. Pascal Sellier. 2004. La sepultura de la cueva Ayayema (Isla Madre de Dios, Archipiélagos Occidentales de Patagonia). *Magallania* 32: 115–124.

Maximiano Castillejo, Alfredo. 2017. From Virtual Survey to Real Prospection: Kawésqar Mobility in the Fuego-Patagonia Seascape across Terrestrial Passages. *Quaternary International* 435: 114–127.

Montenegro, Álvaro, Reneé Hetherington, Michael Eby, and Andrew J. Weaver. 2006. Modelling Pre-Historic Transoceanic Crossings into the Americas. *Quaternary Science Reviews* 25(11–12): 1323–1338.

Orquera, Luis Abel, and Ernesto Luis Piana. 1999. *La vida material y social de los Yámana.* Eudeba, Buenos Aires.

Orquera, Luis Abel, and Ernesto Luis Piana. 2009. Sea Nomads of the Beagle Channel in Southernmost South America: Over Six Thousand Years of Coastal Adaptation and Stability. *Journal of Island and Coastal Archaeology* 4(1): 61–81.

Pallo, María Cecilia. 2011. Condicionamientos de la dinámica ambiental en las decisiones humanas sobre asentamiento y circulación a lo largo del estrecho de Magallanes durante el Holoceno Tardío. *Magallania* 39(2): 177–192.

Perttola, Wesa. 2022. Digital Navigator on the Seas of the Selden Map of China: Sequential Least-Cost Path Analysis Using Dynamic Wind Data. *Journal of Archaeological Method and Theory* 29: 688–721.

Prieto, Alfredo, Denis Chevallay, and David Ovando. 2000. Los pasos de indios en Patagonia Austral. In *Desde el país de los gigantes. Perspectivas arqueológicas en Patagonia,* Tomo 2, edited by Juan Bautista Belardi, Flavia Carballo Marina, and Silvana Espinosa, pp. 87–94. Universidad Nacional de la Patagonia Austral, Río Gallegos.

Reid, Joshua L. 2015. *The Sea Is My Country: The Maritime World of the Makahs.* Yale University Press, New Haven, Connecticut.

Reyes, Omar, César Méndez, Manuel San Roman, Carolina Belmar, and Amalia Nuevo-Delauney. 2022. Biogeographic Barriers in the Circulation and Interaction of Hunter-Gatherer Marine Fishers: The Role of the Taitao Peninsula and the Gulf of Penas (~ 47°S) in the Differentiation of the Cultural Trajectories of West Patagonia. *Frontiers in Earth Science* 10. https://doi.org/10.3389/feart.2022.946732.

Rogers, J. Daniel, and Wendy H. Cegielski. 2017. Building a Better Past with the Help of Agent-Based Modeling. *Proceedings of the National Academy of Sciences of the United States of America* 114(49): 12841–12844.

San Román, Manuel, Omar Reyes, Jimena Torres, and Flavia Morello. 2016. Archaeology of Maritime Hunter-Gatherers from Southernmost Patagonia, South America: Discussing Timing, Changes and Cultural Traditions during the Holocene. In *Marine Ventures: Archaeological Perspectives on Human–Sea Relations,* edited by Hein B. Bjerck, Heidi Mjelva Breivik, Silje E. Fretheim, Ernesto L. Piana, Birgitte Skar, Angélica M. Tivoli, and A. Francisco J. Zangrando, pp. 337–354. Equinox Publishing, Sheffield.

Serrano Montaner, Ramón. 1891. *Derrotero del Estrecho de Magallanes, Tierra del Fuego i canales de la Patagonia: Desde el Canal de Chacao hasta el Cabo de Hornos.* Imprenta Nacional, Santiago de Chile.

Slayton, Emma Ruth. 2018. *Seascape Corridors: Modeling Routes to Connect Communities Across the Caribbean Sea.* Sidestone Press, Leiden.

Smith, Karl. 2020. Modelling Seafaring in Iron Age Atlantic Europe. PhD dissertation, Kellogg College, Oxford University, Oxford.

Stäger, Federico. 2016. *Es un sueño*. Ediciones Una Temporada en Isla Negra, San Antonio, Chile.

Westerdahl, Christer. 2004. On the Significance of Portages—A Survey of a New Research Area. *Nordic Underwater Archaeology*. https://www.abc.se/~m10354/publ/portages.htm, accessed September 11, 2019.

Wilensky, Uri. 1999. NetLogo. Center for Connected Learning and Computer-Based Modeling, Northwestern University, Evanston, Illinois. http://ccl.northwestern.edu/netlogo/.

Zurro, Debora, Virginia Ahedo, María Pereda, Myrian Álvarez, Ivan Briz i Godino, Jorge Caro, José Ignacio Santos, and José Manuel Galán. 2019. Robustness Assessment of the "Cooperation under Resource Pressure" (CURP) Model. *Hunter Gatherer Research* 3(3): 401–428.

3

Seascapes of the Unreal

Using Agent-Based Modeling to Examine Traditional Coast Salish Maritime Mobility

ADAM N. RORABAUGH

Although archaeologists interested in modeling approaches and theory building in general have often referenced Box and Draper's adage "Remember that all models are wrong; the practical question is how wrong do they have to be to not be useful" (1987: 74), a more novel approach is to create "wrong" models that are useful. By this I mean creating intentionally simplified models with the specific intent of making agent-based modeling approaches more accessible to nonexperts while maintaining some utility for answering questions relating to maritime mobility. To that end, the model presented here is a simple model of maritime mobility expanding on the effort by Ames (2002) in "Going by Boat: The Forager-Collector Continuum at Sea" and integrating a geographic information system (GIS) with Unreal Engine 5.2.1 (UE5). Unreal Engine is a free-to-download toolset originally designed for game development but with broader applications in architecture and media production and designed to have a low barrier of entry for those without programming language experience.

Generally, simple mathematical or agent-based models enable archaeologists to test the assumptions of theoretical narratives by creating cultural laboratories (Blankshein 2022; Callaghan and Scarre 2009; Porčić 2023; see the chapters in Romanowska et al. 2021). In the case of maritime travel, does water transportation represent a low-cost mode of travel (Ames 2002; Gustas and Supernant 2017; Ritchie and Miller 2021; Rorabaugh 2015, 2019) that can be represented with aspatial network models or do other factors weigh distances, contribute risk and cost, or add a temporal dimension to variability (see Garcia-Piquer, this volume: Chapter 2)? The implications of such simple models can have drastic impacts on archaeological interpretations and on empirical sufficiency regarding the archaeological record (Barrett 2019; Deffner et al. 2022; Rorabaugh 2014).

Additionally, the focus of this model using Unreal Engine is for pedagogical reasons. Traditionally, agent-based models have required extensive programming experience, and even the more approachable introductory methods for agent-based modeling such as the NetLogo programming language have a high barrier of entry for STEM teachers (Borowczak and Burrows 2019). Other approaches require familiarity with R, C++, and Python, which may not be accessible for many anthropology students even at the graduate level. In the words of Chevy Chase's Gerald R. Ford, "It was my understanding that there would be no math." The focus on using game development tools for teaching archaeological modeling and advancing pedagogies of land (e.g., Styres et al. 2013; Tuck et al. 2014) is to make agent-based modeling in archaeology less intimidating and to demonstrate its broader benefit.

The use of gaming and game development tools for advancing archaeological and place-based learning is but one aspect of what Reinhard (2018) defined as archaeogaming. More broadly, archaeogaming is the study of archaeology in, and of, tabletop games, video games, and their media representations. It is the examination of said media using archaeological perspectives and then critically examining the portrayal of archaeological ethics in these media. The use of a game to examine archaeological questions may be best viewed as "experimental archaeogaming" in the same vein that producing traditional lithic tools and examining their debitage or tool use and wear characteristics is experimental archaeology. The aim of this experimental archaeogaming model is to help advance place-based learning (Chiblow and Meighan 2022) as a complementary approach to ArcGIS StoryMaps with place names (e.g., Palmer-McGee 2022). This chapter aims to provide a proof-of-concept model using ethnographic data and Unreal Engine to demonstrate the utility of using tools intended for game development to lower the barrier of entry for the agent-based modeling of anthropological and archaeological questions.

Background

Traditional Mobility in the Salish Sea

The watery world examined in this study is the Salish Sea, composed of the Puget Sound and Strait of Georgia and their respective watersheds, with a specific focus on the traditional maritime territories of the Lummi, Nooksack, Samish, Stillaguamish, Semiahmoo, Lower Skagit, and Suquamish Coast Salish communities. Archaeologists have long recognized the importance of mobility in the patterning of the archaeological record, particularly with complex sedentary foragers (Binford 1980; Kelly 1983, 1992, 2007).

Maritime mobility fundamentally patterns the traditional seasonal mobility strategies of Coast Salish peoples, which have been well documented in the ethnohistoric record and oral traditions (Ames and Maschner 1999; Hill-Tout 1978; Kelly 1983: 280; Mitchell 1979; Moss 2012; Suttles 1990; Teit 1928). Two to four residential moves were typical per year (Kelly 1983: 280), with large, aggregated villages composed of multiple households or kin groups during the fall and winter months (Barnett 1955; Hill-Tout 1978; Mitchell 1979; Suttles 1960, 1974, 1990; Teit 1928). Seasonal resources including, but not limited to, geophytes, anadromous fish, terrestrial or marine mammals, and marine invertebrates were mass harvested by these aggregated groups for winter storage.

Over the spring and summer months these villages dispersed into smaller household groups for harvesting seasonal resources. Resources were owned by extended kin groups, which also managed resource access (Suttles 1960). Bilateral descent (Suttles 1960, 1974), the use of boats for long-distance travel (Ames 2002), and ownership of resources by extended kin groups resulted in expansive and highly variable traditional social networks for people in the precontact Coast Salish world.

Experimental Archaeogaming

Archaeogaming is the critical examination of how archaeology as a discipline is represented in media (Reinhard 2018). Experimental archaeogaming is the use of games or game development tools to answer or interpret material culture. In this case, the use of UE5, a game development tool, as opposed to more traditional programming languages (Romanowska et al. 2021) or software widely utilized (Borowczak and Burrows 2019) in archaeological and biological agent-based modeling, makes the presented model a type of experimental archaeogaming.

Although social science undergraduate students often have considerable literacy in interactive media, and some even have a degree of programming and data management familiarity from the toolkits to modify games (Reinhard 2018), math and statistical anxiety are significant barriers for undergraduate data literacy in the social sciences (Condon et al. 2023). Based on such studies and cursory searches of undergraduate and graduate program requirements for statistics in anthropology programs in North America and the United Kingdom, it is not controversial to state that many students enter anthropology and other social sciences due to the lower barrier of entry in terms of statistics, programming, and mathematics compared to many other disciplines. As the discipline moves toward less extractive research methods using big data (see

Figure 3.1. Example of Unreal Engine's blueprint interface.

Garcia-Piquer, this volume: Chapter 2; and Jarrett, this volume: Chapter 12), and as the clear needs of the discipline to increase meaningful diversity, equity, and inclusion (e.g., Gamble et al. 2020), what approaches can be taken to enable inclusivity in big data and modeling approaches and provide potential researchers the tools to succeed in the face of anxiety to math, statistics, and programming?

One possible solution is the use of visual scripting in free-to-access software. To that end, UE5 was used with its "blueprint" visual scripting system (Figure 3.1). The Unreal Editor software is free to download for potential developers, and the project file can be accessed and modified by anyone with Unreal Editor. One advantage with UE5 over other visual scripting platforms for archaeological studies is the support for integration with ESRI GIS software.

Methodology

Modeling Traditional Straits Salish Maritime Mobility

The main question of this model is exploratory: what patterns emerge when applying seagoing canoe speeds and mobility patterns to locations not discussed in ethnographic accounts? The model constructed in UE5 consists of a landscape based on the ESRI plugin for Unreal Engine using a 10 m digital elevation map. Agents take the form of Unreal Engine "actors." The model consisted of two types of actors: site locations (Figure 3.2) and sea canoes (Figure 3.3).

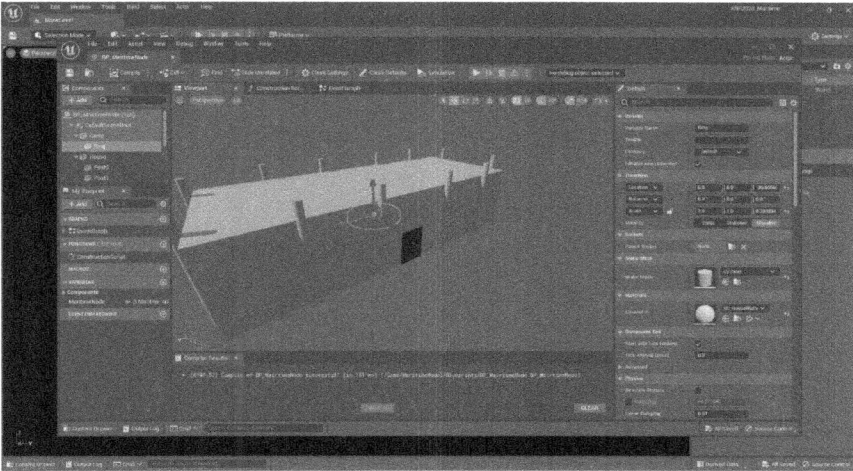

Figure 3.2. Site location Unreal actor.

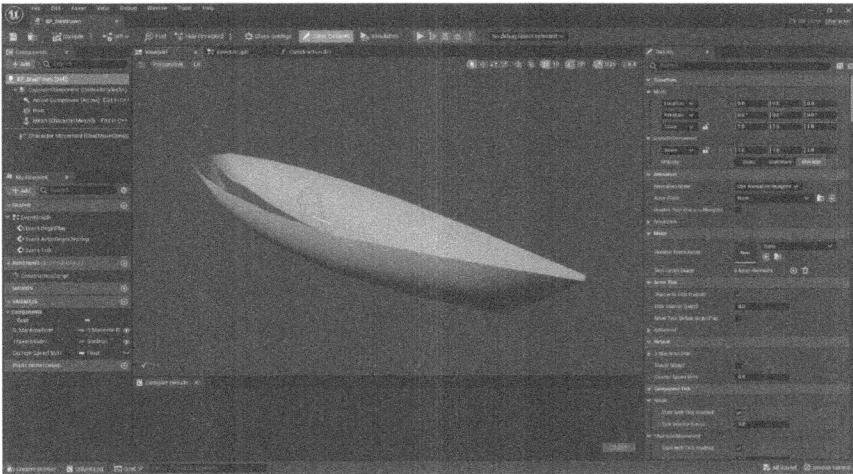

Figure 3.3. Sea canoe Unreal actor.

Site Locations

Site locations were divided into residential sites or activity camps based on whether there was an observed archaeological or ethnographic expression of a residential site component. This is an abstraction as many sites in the Salish Sea have multiple components with functional uses that change through time, although there is continuity in use of place (e.g., Brown 2022; Hopt and Grier 2017; Morin et al. 2018), and some sites have been subject to substantial erosion or impacts from development. To protect site locations, points for the site actors are based on 100 m buffered random locations on the landscape. Out of an examined sample of 35 modeled site locations, 25 were classified as residential sites and 10 as activity camps (Figure 3.4). Activity camps were limited to a sample of Straits Salish activity camps, specifically eastern Rosario Strait, while residential locations were drawn from a larger archaeological sample to simulate long-distance movement.

Sea Canoes

The watergoing craft chosen as agents in this agent-based model were sea canoes. With few exceptions, canoes in the Salish Sea were dugouts from western red cedar logs (Boas 1909; Drucker 1951). There was considerable variation in the types of watercraft traditionally used. The largest vessels, "great canoes," were used for warfare and freight. Great canoes were over 10 m long with high sides and were broad beamed (Drucker 1963), but 18 m was the practical size limit and 12 m an average length. Freight canoes were considered slow and cumbersome but used to transport household goods in fair weather conditions, and house planks could be placed across two tied canoes to support additional cargo (Drucker 1951: 88). Great canoes used for warfare and as seagoing vessels tended to have beams of 2 m and were narrower than bespoke freight canoes, and some vessels were specifically constructed for conflict. In Puget Sound, "household canoes" would travel in groups and carry around nine individuals per canoe to potlatches (Castile 1985). Medium-sized "family canoes" with lengths of 5.5–11 m and 1–2 m beams were also used, along with smaller vessels for one or two people (Drucker 1963). For this model, canoes with 10 m length and 2 m beam were used, falling under the household and larger family canoe category as opposed to great canoes.

Modeling Travel

Two travel modes were modeled: (1) local mode, involving transport from a residential location to an activity location, and (2) travel mode, which was residential site to residential site. Logistically for the purposes of trip planning, this dichotomy reflects the reality that individuals would be prepared for trips to other residential locations or trips to resource collection areas. For ease of interpretation and programming, this model is limited to point-to-point travel. Traditionally, individuals or corporate groups may have traveled to another residential site to gain permission to access local resources for forays from there and return to the residential site of origin later.

Speeds for boat travel were given minimum and maximum ranges based on transport mode and based on estimates derived from past studies and eth-

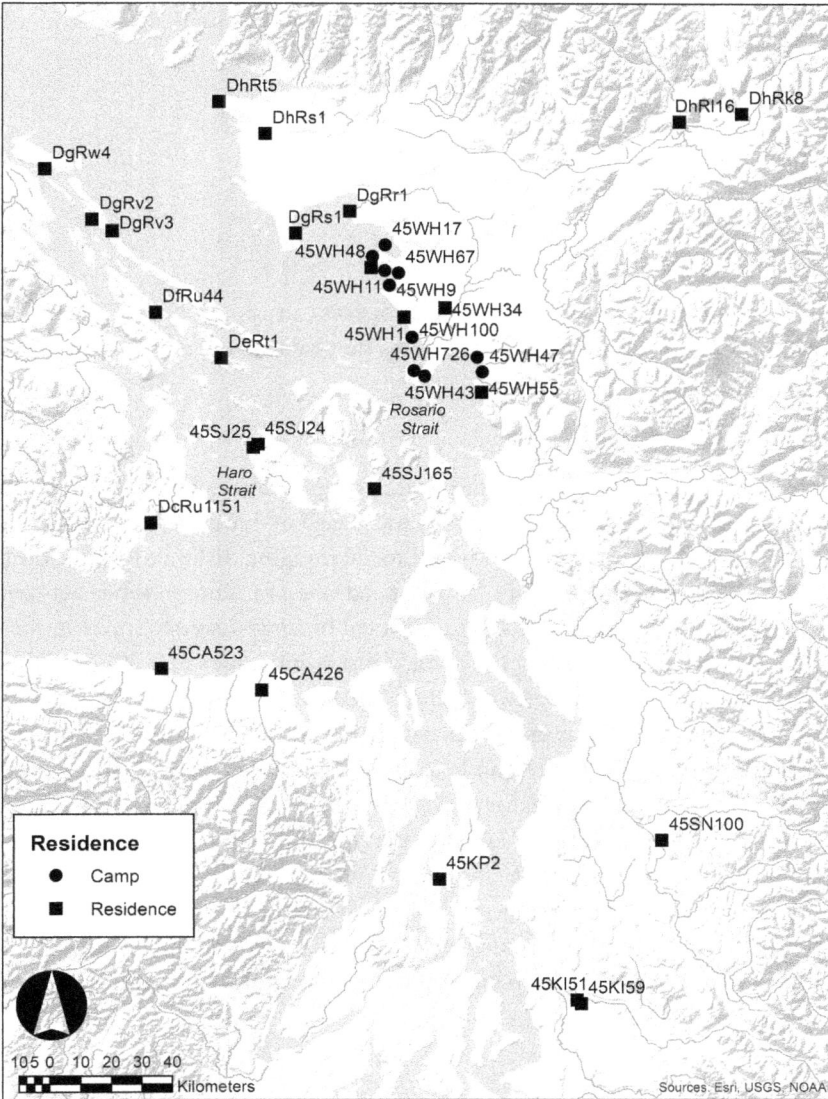

Figure 3.4. Salish Sea sites in the model.

nographic accounts (e.g., Ames 2002; Castile 1985; Croes and Hackenberger 1988; Durham 1960; Sproat 1987 [1868]; Swan 1967 [1857]). The variability in reported speed is a combination of different vessel loads, crew skill, and a range of weather and current conditions. As the data derived from the ethnographic accounts is limited to recorded voyages between specific locations, this model

Table 3.1. Model Parameters by Travel Mode

Parameter	Local	Long Distance
Minimum Speed (km/h)	2.7	4.4
Maximum Speed (km/h)	4.5	6.5

Source: Ames 2002: 30–31.

allows for a projection onto other locations not in the ethnographic record while examining variability in travel times that captures a range of skill levels and conditions.

Sample and Model Parameters

In the sample, there were 660 valid point-to-point dyads between sites. The division between short and long forays was based on whether the destination was within a radius of 4.4 km, the noted social foraging distance for the Straits Salish (Schalk 1978: Table 4–11). Travel speed was a random number between the minimum and maximum values for local or long-distance travel modes. Agents traveled on a least-cost path (e.g., Gustas and Supernant 2017; Mlekuž 2012) on an artificial intelligence (AI) navigation mesh to other site locations as opposed to geodesic distance. AI navigation meshes are essentially a grid with weighted areas for more and less likely transit, with barriers being areas that the AI will not have an actor move through. In the case of this model, the barriers are land. Conditions such as currents and weather were abstracted into the travel speed variability (Table 3.1) and acted as weight on the least-cost path on the navigation mesh. The model had 500 runs, and the travel time for each least-cost path was exported into a comma delineated table. A total of 330,000 journeys between sites were simulated. The results were then analyzed in R 4.3.0.

Using and Adjusting the Model

The model is available through the Network for Computational Modeling in the Social and Ecological Sciences (CoMSES Net) and can be loaded as an unreal project in UE5 with the ESRI plugin installed. The project will load the model's level by default (Figure 3.5). Pressing "play" in the editor runs the model with a text output in the editor's log that can be exported as comma delineated data for analysis. Model parameters and data can be edited by adjusting data tables or editing the actors in the editor.

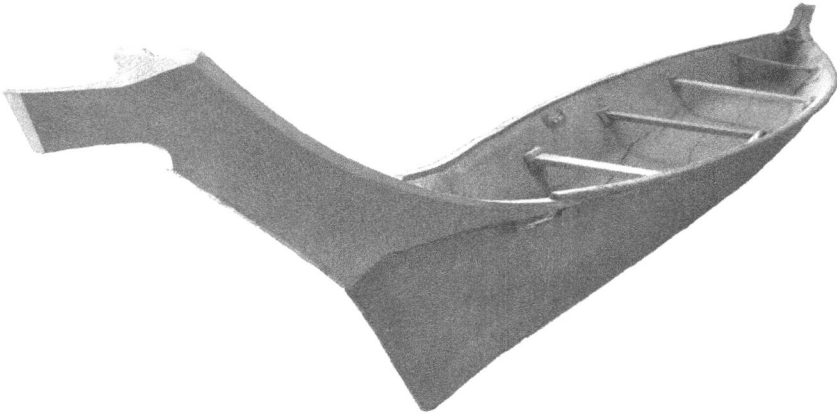

Figure 3.5. Coast Salish "family canoes" like this one, currently exhibited at the Museum of Anthropology of Vancouver (BC, Canada), were used as an inspiration to the 3-D model used in the simulation. Photo by Alberto García-Piquer.

Figure 3.6. Screenshot of Unreal Editor with a model loaded.

Results

Watercraft speeds and travel times were examined by travel modes. The difference in watercraft speeds between local mode and long-distance modes (Figure 3.6) was found to be highly statistically significant (Mann-Whitney U = 14.4, p <0.001, N = 330,000). Median speeds were 5.5 km/h for long-distance mode and 3.5 km/h for local mode. While intuitively these results may appear

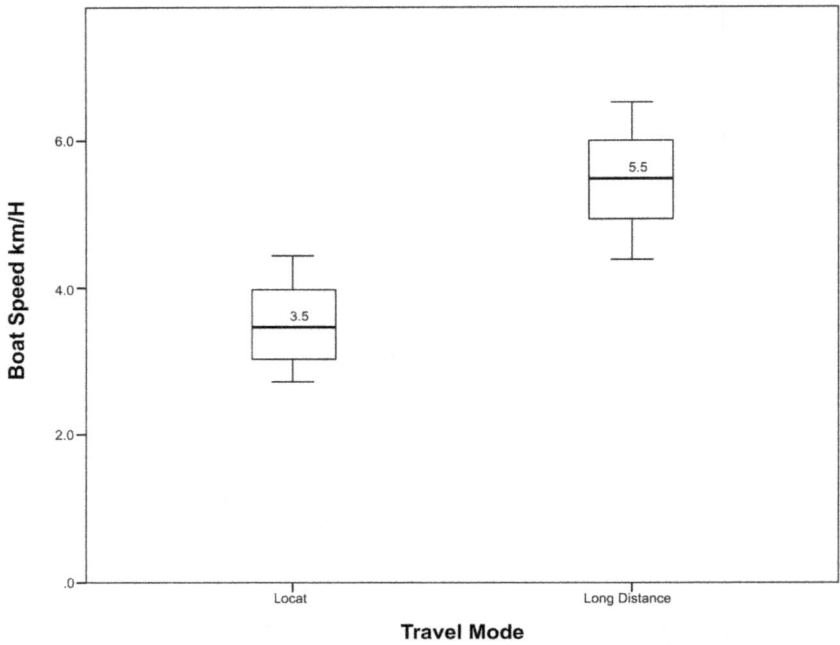

Figure 3.7. Whisker plot of boat speeds by travel mode (*N* = 330,000).

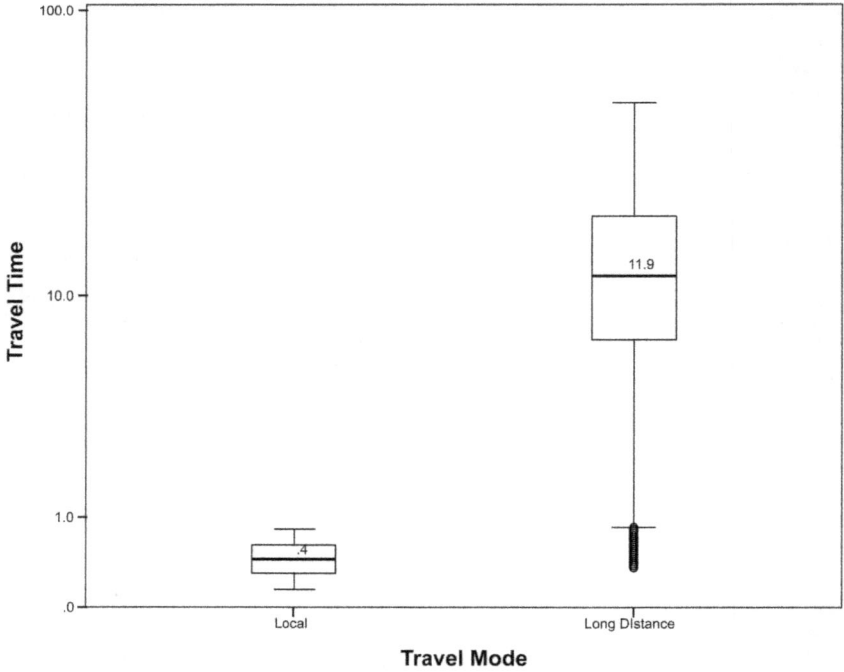

Figure 3.8. Whisker plot of travel time by travel mode (*N* = 330,000).

to have been the case from ethnohistoric sources discussing maritime travel, by including a larger sample of sites and assessing variation in a simple model, the strength of these patterns can be further explored.

Discussion

This experimental archaeogaming model expanded on past studies by modeling variation in travel speeds, abstracting skill and conditions to documented archaeological sites in the Salish Sea, focusing on the Straits Salish. Trips tended to be short half- to full-day forays when accounting for gathering activities or longer journeys to residential sites. Most residential locations in the Salish Sea were reachable by sea canoe within two or three days. While this model suggests that overall transportation costs to reach sites may be low, variability in transport risk and cost cannot be dismissed.

Although Leary (2014: 6) has argued that mobility lacks a presence in the archaeological record, modeling approaches such as the one presented here provide methodological tools to examine traditional mobility. Similarly, the model presented here by using an AI navigation mesh expands on past efforts to examine mobility that have emphasized least-cost-path analysis (Gustas and Supernant 2017; Mlekuž 2012; Verhagen et al. 2019). Seascapes are not as fixed, tangible, or "textured" as landscapes (Blankshein 2022). This issue can be engaged at different levels of abstraction, such as the high degree in the simple model provided here, or more direct engagements (see Smith, this volume: Chapter 9). Additionally, the Salish Sea provides a direct contrast to other regions where there is limited evidence of the watercraft used (Blankshein 2022) as there is a rich record to draw on for ethnographically informed questions (e.g., Grier and Shaver 2008) to develop models and refine our questions.

The model presented here is a proof of concept for using game development tools for agent-based modeling and is readily expandable. Areas for future work include, but are not limited to, other types of craft such as river canoes and cargo canoes or examining variation in the social foraging ranges of other Coast Salish groups and its impact on travel times. Special but frequent uses of canoes such as side-by-side canoes for cargo or house plank transport can also be considered in more bespoke models for specific traditional transportation questions. Decoupling travel risk factors from travel speed variability by constructing a riskscape to weight the AI navigation mesh is another potential direction. Such a riskscape could weigh factors such as underlying bathymetry, currents, tides, and weather risk (see Garcia-Piquer, this volume: Chapter 2) in cells used for AI travel assessment. Different seasonal riskscapes could be used to assess seasonal transport variation. In a similar vein, "sea truthing" models such as this with community engagement (see Rivera Prince, this volume: Chapter 11) is another potentially fruitful direction to refine maritime mobility models.

Simple experimental archaeogaming models such as the one presented here can provide new means of engaging with the archaeological, ethnographic, and oral records. The increasing availability and accessibility of programming and visualization tools such as Unreal Engine also provide opportunities to reduce the barrier of entry for descendant communities, students, and professional archaeologists to engage with the past in new ways. By reducing these barriers, the utility of applying modeling approaches to the past can be better realized, and archaeologists engaged in agent-based models can meet the call to increase representation (e.g., Gamble et al. 2020).

This approach constitutes a non-extractive form of archaeology, and while it can in no way replace community place-based learning (see Chiblow and Meighan 2022; Styres et al. 2013; Tuck et al. 2014), it and virtual places like it can provide engagement with landscapes and places that no longer exist or are utterly unrecognizable due to development or anthropogenic climate change. Such digital storytelling can be complementary to engaging with the past through other geospatial approaches like StoryMaps (e.g., Palmer-McGee 2022).

Model Availability

Rorabaugh, Adam. 2023, December 22. Seascapes of the Unreal: Using Agent Based Modeling to Examine Traditional Coast Salish Maritime Mobility (Version 0.0.1). CoMSES Computational Model Library. Retrieved from: https://www .comses.net/codebases/bcaaea5f-8552-4427-90c3-edceee6047aa/releases/0.0.1/

Acknowledgments

I thank Elliot Helmer, Anna Coon, and Dennis Lewarch for their thoughtful insights and suggested literature at different stages of this project, and the editors of this volume.

References Cited

Ames, Kenneth. 2002. Going by Boat: The Forager-Collector Continuum at Sea. In *Beyond Foraging and Collecting: Evolutionary Change in Hunter-Gatherer Settlement Systems,* edited by Ben Fitzhugh and Junko Habu, pp. 17–50. Kluwer / Plenum Press, New York.

Ames, Kenneth, and Herbert D. G. Maschner. 1999. *Peoples of the Northwest Coast: Their Archaeology and Prehistory.* Thames and Hudson, London.

Barnett, Homer G. 1955. *The Coast Salish of British Columbia.* Studies in Anthropology 4. University of Oregon Press, Eugene.

Barrett, Brendan J. 2019. Equifinality in Empirical Studies of Cultural Transmission. *Behavioural Processes* 161: 129–139.

Binford, Lewis R. 1980. Willow Smoke and Dogs' Tails: Hunter-Gatherer Settlement Systems and Archaeological Site Formation. *American Antiquity* 45: 4–20.

Blankshein, Stephanie L. 2022. (Sea)ways of Perception: An Integrated Maritime-Terrestrial Approach to Modelling Prehistoric Seafaring. *Journal of Archaeological Method and Theory* 29: 723–761.

Boas, Franz. 1909. *The Ethnology of the Kwakiutl.* Memoirs of the American Museum of Natural History Vol. 8 pt 2. Government Printing Office, Washington, D.C.

Borowczak, Mike, and Andrea C. Burrows. 2019. Ants Go Marching: Integrating Computer Science into Teacher Professional Development with NetLogo. *Education Sciences* 9: 66.

Box, George E. P., and Norman R. Draper. 1987. *Empirical Model Building and Response Surfaces.* Wiley, New York.

Brown, James W. 2022. Creating Neighborhoods: Cultural, Spatial, and Temporal Evaluation of Large Shell-Bering Sites in the San Juan Islands of Washington State. Paper presented at the 87th Society for American Archaeology Annual Meeting, Chicago, Illinois.

Callaghan, Richard, and Chris Scarre. 2009. Simulating the Western Seaways. *Oxford Journal of Archaeology* 28(4): 357–372.

Castile, George P. (editor). 1985. *The Indians of Puget Sound: The Notebooks of Myron Ells.* University of Washington Press, Seattle.

Charles, S. [suetv] 2019. Tribal Journeys Info. Journey 2019 Map. Electronic document, https://tribaljourneys.wordpress.com/2019/07/08/journey-2019-map, accessed May 26, 2023.

Chiblow, Susan, and Paul J. Meighan. 2022. Language Is Land, Land Is Language: The Importance of Indigenous Languages. *Human Geography* 15(2): 206–210.

Condon, Patricia B., Elita Exline, and Louise Buckley. 2023. Data Literacy in the Social Sciences: Findings from a Local Study on Teaching with Quantitative Data in Undergraduate Courses. *Evidence Based Library and Information Practice* 18(1): 61–75.

Croes, Dale, and Steven Hackenberger. 1988. Hoko River Archaeological Complex: Modeling Prehistoric Northwest Coast Economic Evolution. *Research in Economic Anthropology Supplement* 3: 19–86.

Deffner, Dominik, Anne Kandler, and Laurel Fogarty. 2022. Effective Population Size for Culturally Evolving Traits. *PLOS Computational Biology* 18(4): e1009430.

Drucker, P. 1951. *The Northern and Central Nootkan Tribes.* Bureau of American Ethnology Bulletin 144. Smithsonian Institution, Washington, D.C.

Drucker, Philip. 1963. *Indians of the Northwest Coast.* Natural History Press, Garden City, New York.

Durham, Bill. 1960. *Canoes and Kayaks of Western North America.* Copper Canoe Press, Seattle.

Gamble, Lynn, Debra Martin, Julia Hendron, Calogero Santoro, Sarah Herr, Christina Rieth, Sjoerd van der Linde, Christopher Rodning, Michelle Hegmon, and Jennifer

Birch. 2020. Statement and Commitments from SAA Editors to Change the Under-representation of Black, Indigenous, and Other Scholars from Diverse Backgrounds in Our Publications. Electronic document, https://www.saa.org/quick-nav/saa-media-room/saa-news/2020/07/01/statement-and-commitments-from-saa-editors-to-change-underrepresentation, accessed June 26, 2024.

Grier, Colin, and Chief Lisa Shaver. 2008. The Role of Archaeologists and First Nations in Sorting Out Some Very Old Problems in British Columbia. *SAA Record* 8(1): 33–35.

Gustas, Robert, and Kisha Supernant. 2017. Least Cost Path Analysis of Early Maritime Movement on the Pacific Northwest Coast. *Journal of Archaeological Sciences* 78: 40–56.

Hill-Tout, C. 1978. *The Salish People. 3: The Mainland Halkomelem.* Talon Books, Vancouver, British Columbia.

Hopt J., and C. Grier. 2017. Continuity Amidst Change: Village Organization and Fishing Subsistence at the Dionisio Point Locality in Coastal Southern British Columbia. *Journal of Island and Coastal Archaeology* 13(1): 21–42.

Kelly, Robert L. 1983. Hunter-Gatherer Mobility Strategies. *Journal of Anthropological Research* 39(3): 277–306.

Kelly, Robert L. 1992. Mobility/Sedentism: Concepts, Archaeological Measures, and Effects. *Annual Review of Anthropology* 21: 43–66.

Kelly, Robert L. 2007. *The Foraging Spectrum: Diversity in Hunter-Gatherer Lifeways.* Percheron Press, New York.

Leary, Jim. 2014. Past Mobility: An Introduction. In *Past Mobilities: Archaeological Approaches to Movement and Mobility,* edited by Jim Leary, pp. 1–19. Routledge, New York.

Mitchell, Donald H. 1979. Seasonal Settlements, Village Aggregations, and Political Autonomy on the Central Northwest Coast. In *The Development of Political Organization in Native North America.* Proceedings of the American Ethnological Society, edited by Elisabeth Tooker and Morton H. Fried, pp. 97–107. American Ethnological Society, Washington, D.C.

Mlekuž, Dimitrij. 2012. Everything Flows: Computational Approaches to Fluid Landscapes. In *Archaeology in the Digital Era: Papers from the 40th Annual Conference of Computer Applications and Quantitative Methods in Archaeology (CAA), Southampton, 26–29 March 2012,* vol. 1, pp. 839–845. Amsterdam University Press, Amsterdam.

Morin, Jesse, Dana Lepofsky, Morgan Ritchie, Marko Porčić, and Kevan Edinborough. 2018. Assessing Continuity in the Ancestral Territory of the Tsleil-Waututh Coast Salish, Southwest British Columbia, Canada. *Journal of Anthropological Archaeology* 51: 77–87.

Moss, Madonna L. 2012. *Northwest Coast: Archaeology as Deep History.* SAA Press, Washington, D.C.

Palmer-McGee, Casey. 2022. Coast Salish Place Names of the San Juan Islands. Samish Tribe Electronic document, https://storymaps.arcgis.com/stories/9b0f86b51e054ba78b83ab39c4d0b1a6, accessed July 4, 2023.

Porčić, Marko. 2023. *Patterns in Space and Time: Simulating Cultural Transmission in Archaeology.* Laboratory for Bioarchaeology, Faculty of Philosophy, University of Belgrade, Belgrade.

Reinhard, Andrew. 2018. *Archaeogaming: An Introduction to Archaeology in and of Video Games.* Berghahn, New York.

Ritchie, Morgan, and Bruce Granville Miller. 2021. Social Networks and Stratagems of Nineteenth-Century Coast Salish Leaders. *Ethnohistory* 68(2): 237–268.

Romanowska, Iza, Colin D. Wren, and Stefani A. Crabtree. 2021. *Agent-Based Modeling for Archaeology: Simulating the Complexity of Societies.* Sante Fe Institute Press, Santa Fe, New Mexico.

Rorabaugh, Adam N. 2014. Impacts of Drift and Population Bottlenecks on the Cultural Transmission of a Neutral Continuous Trait: An Agent-Based Model. *Journal of Archaeological Science* 49: 255–264.

Rorabaugh, Adam N. 2015. Modeling Pre-European Contact Coast Salish Seasonal Social Networks and Their Impacts on Unbiased Cultural Transmission. *Journal of Artificial Societies and Social Simulation* 18(4): 8.

Rorabaugh, Adam N. 2019. Hunting Social Networks on the Salish Sea Before and After the Bow and Arrow. *Journal of Archaeological Science: Reports* 23: 822–843.

Schalk, Randall F. 1978. Foragers of the Northwest Coast of North America: The Ecology of Aboriginal Land Use Systems. PhD dissertation, Department of Anthropology, University of New Mexico, Albuquerque.

Sproat, Gilbert Malcolm. 1987 [1868]. *The Nootka: Scenes and Studies of Savage Life.* Sono Nils Press, Victoria.

Styres, Sandra, Celia Haig-Brown, and Melissa Blimkie. 2013. Toward a Pedagogy of Land: The Urban Context. *Canadian Journal of Education / Revue Canadienne de l'éducation* 36(2): 34–67.

Suttles, Wayne P. 1960. Affinal Ties, Subsistence, and Prestige among the Coast Salish. *American Anthropologist* 62: 296–305.

Suttles, Wayne P. 1974. *The Economic Life of the Coast Salish of Haro and Rosario Straits.* Coast Salish and Western Washington Indians. Garland, New York.

Suttles, Wayne P. 1990. Central Coast Salish. In *Northwest Coast,* edited by Wayne P. Suttles, pp. 453–475. Vol. 7 of *Handbook of North American Indians,* William C. Sturtevant, general editor. Smithsonian Institution, Washington, D.C.

Swan, James Gilchrist. 1967 [1857]. *The Northwest Coast.* University of Washington Press, Seattle.

Teit, James H. 1928. *The Middle Columbia Salish.* University of Washington Publications in Anthropology Vol. 2, No. 4. University of Washington Press, Seattle.

Tuck, Eve, Marcia McKenzie, and Kate McCoy. 2014. Land Education: Indigenous, Post-Colonial, and Decolonizing Perspectives on Place and Environmental Education Research. *Environmental Education Research* 20(1): 1–23.

Verhagen, Philip, Laure Nuninger, and Mark R. Groenhuijzen. 2019. Modelling of Pathways and Movement Networks in Archaeology: An Overview of Current Approaches. In *Finding the Limits of the Limes: Modelling Demography, Economy and Transport on the Edge of the Roman Empire,* edited by Philip Verhagen, Jamie Joyce, and Mark R. Groenhuijzen, pp. 217–249. Springer, Cham.

4

Were Sperm Whales Hunted by Megalithic Communities in Brittany, France, During the Fifth Millennium cal BCE?

BETTINA SCHULZ PAULSSON

Megalithic art in Brittany, France, serves as evidence of the maritime connection of coastal communities during the fifth millennium cal BCE. Iconographic representations found on standing stones and within megalithic tombs featuring motifs such as boats and sperm whales offer insights into the maritime technologies of these early megalithic societies.

The engravings of the sperm whales have been the subject of much discussion, with theories proposing that they might result from observations of stranded animals or from witnessing the whales in their natural habitat. Limited consideration has previously been given to the possibility of active whaling on the open ocean.

The primary objective of this chapter is to offer a contribution to the ongoing debate concerning whether these coastal Stone Age communities could possess the capacity for whale hunting. It provides evidence for this practice through comparison to whaling rock art in regions such as Fennoscandia and Chile, dating back to the sixth and the fifth millennia cal BCE. This rock art emphasizes the importance of whaling in the subsistence of coastal Stone Age communities in the middle Holocene and suggests that the hunting strategies were highly sophisticated and required a well-organized social structure.

The Prehistoric Exploitation of Large Whales: Active Hunting or Scavenging?

Marine mammal exploitation has been part of the subsistence of coastal groups since earliest times and predates modern humans. Caves across Gibraltar hold evidence for the harvest of sea mammals in the Mediterranean as early as the

Middle Paleolithic at sites attributable to the Mousterian and the Neanderthals (Colonese et al. 2011; Erlandson 2001; Stringer et al. 2008).

For northern Europe, there is a large body of evidence for the prehistoric exploitation of pinnipeds (seals) and cetaceans (dolphins, porpoises, whales) (Storå and Lõugas 2005). The presence of cetacean bones as an indicative sign of whaling at archaeological sites along the Atlantic shores of northern Europe is well documented (e.g., Clark 1947; Erlandson 2001; Herman and Dobney 2004). It is apparent that large whale species were subject to early exploitation. For instance, evidence from the Neolithic Orkney Islands and the Outer Hebrides suggests the harvesting of humpback and mink whales, with Bronze Age contexts on these islands containing the bones of sperm whales (Buckley et al. 2014: Table 4).

However, there is an old and long-standing debate as to whether large whales were obtained as an occasional windfall due to natural stranding or whether they were hunted actively on the open ocean, as discussed in prior studies (e.g., Castelleti 2020; Clark 1947; Erlandson 2001; Mulville 2002; Pétillon 2016). Evidence of active hunting or slaughtering of cetaceans is infrequent and primarily associated with smaller species such as dolphins. One notable instance involves the Mesolithic period where the hunting of white-nosed dolphins is supported by traces on carcasses discovered in Huseby Klev on the Swedish island of Orust (Boethius 2018). The dolphins were corralled into a bay and subsequently slaughtered.

The prehistoric hunting of large whale species, such as sperm whales, is presumed to have had a limited impact on the skeletal apparatus of these animals, resulting in only a few discernible traces. In addition, prehistoric cetacean bones typically exhibit a relative fragility attributed to their diminished mineral content. This is why many archaeological specimens are preserved in a fragmented state, which makes it difficult to determine the species. Cetacean bones in collections are often referred to only as marine mammal bones (Buckley et al. 2014). Another reason is that bones and baleens were often reworked and used as building material, tools, and ornaments (e.g., Charpentier et al. 2022; Pétillon 2016; Schuhmacher et al. 2013; Sinding et al. 2012).

The earliest written references to the active hunting of marine mammals can be traced back to early Christian texts from the seventh and eighth centuries. Bede, in his work from AD 731, mentions the hunting of both seals and cetaceans. Subsequently, there are accounts of herding, stranding, and the slaughter of small whales in the western and northern isles of Scotland and in Iceland (Fenton 1997; Kristjánsson 1986). These historical texts lack comprehensive details regarding the specific hunting and whaling strategies employed in those times (Szabo 2008).

In recent years a burgeoning body of evidence has emerged revealing an increasing number of sperm whale engravings within early megalithic contexts in the Morbihan region of Brittany (Cassen and Vaquero 2000; Cassen et al. 2021; Cassen and Grimaud 2020a, 2020b; Whittle 2000). These engravings distinctly portray direct observations of these marine mammals in their natural environment (Cassen and Vaquero 2000). The emergence of recent findings that illustrate the coexistence of sperm whales and small boats prompts inquiries into potential whaling practices. Were these early megalithic societies hunting whales?

When the archaeological evidence is sparse, rock art is one of the best kinds of evidence that can provide insights into the strategies of exploitation of marine mammals and maritime technologies. Stone Age societies in the middle Holocene were whaling, as we showcase with whaling rock art in Fennoscandia and the Atacama Desert.

Megalithic Sperm Whales and Early Coastal Stone Age Societies (4700–3500 cal BCE)

Brittany, and Morbihan in particular, is the region with the earliest megalithic structures in Europe (Schulz Paulsson 2017, 2019) and the region with the earliest and most complex megalithic art. Engravings have been documented in at least 83 sites, including standing stones, alignments, and megalithic tombs (Schulz Paulsson 2025).

Brittany is also the only megalithic region in Europe with rock art that depicts seascapes with distinct maritime fauna (Cassen and Vaquero 2000; Cassen et al. 2021; Cassen and Grimaud 2020a, 2020b). It is also the region with the earliest and the most frequent depictions of megalithic boats (Schulz Paulsson 2023).

Over the past two decades there has been a notable shift in the terrestrial interpretation of certain symbols, with a growing inclination toward accepting a maritime understanding. The most significant shift involves the depictions of sperm whales. Initially, these symbols were construed as representations of plows or as Mané Rutual–type axe (Shee Twohig 1981), but they have since been reevaluated and identified as depictions of sperm whales, a theory first proposed by Alasdair Whittle (2000) and Serge Cassen and Jacobo Vaquero (2000). While early portrayals of sperm whales were abstract and stylistic, the last decade has witnessed the identification of an increasing number of clear and distinct depictions, especially on the standing stones Kermaillard, Saint Samson, and in the megalithic grave Mané Lud. In recent years the megalithic art of Brittany has been the subject of the extensive research program Corpus des signes gravés néolithiques, Programme collectif de recherche (PCR), within

Figure 4.1. Sperm whale engravings in megalithic Brittany. *Left*: Back orthostat in the passage grave Mané Lud, Locmariaquer: *top,* photo enhanced with YRE algorithm; *bottom,* laser scan visualized with the software Polyview and analyzed with different light angles, interpreted in white. *Right:* Photo of a capstone reused standing stone with sperm whale engraved on the inner part, interpreted in white.

the framework of which it has been possible to identify not only more sperm whales but also the images of giant squids (Cassen et al. 2021).

The number of sperm whale engravings has been increasing in the last few years, and we now know of whale depictions from around twelve sites. These whales are depicted with some common anatomical features. The head part is thickened, the flukes are in motion, and the blow is clearly shown in the form of a bow above the head (Figure 4.1). Some of the animals are depicted with their sex discernible (in Kerdual, Pen Harp, and Cruguellic). It is essential to comprehend that these engravings portray living animals and observations of sperm whales in their natural environment and the open sea (Cassen and Vaquero 2000). Sperm whales congregate along the continental shelf, where they dive deep into the sea in pursuit of their preferred prey, giant squids. This occurs approximately 200 km beyond the Bay of Morbihan or about 150 km from the western coast of Île d'Yeu (western France), which represents the shortest distance from the land to the continental shelf.

The sperm whales are either engraved on standing stones or on reused standing stones in passage graves and are associated with the earliest megalithic phases in Brittany, the ancient Castellic horizon (4700–4200 cal BCE). Symbolic expression in these early megalithic contexts begins with complex symbolic imagery featuring iconic signs, symbols, and combinations of symbols. The earliest main signs include boats, whales, zigzags/waves, jadeite axes, blades, and crooks. Interesting for our discussion is the fact that these whales are often combined with a square, which could indicate a region or hunting territory.

There is limited evidence of whale bones from Stone Age contexts in Brittany. The region's soil is highly acidic, resulting in generally poor bone preservation. Bones tend to be only conserved in calcareous environments like shell banks. It may be added that, as already mentioned at the outset, the multifunctional use of whale bones subjects them to fragmentation. In a few cases from Brittany and adjacent regions, however, whale bones have been discovered. For instance, on the islands of Hoëdic and Téviec are Mesolithic graveyards dating back to the seventh to the fifth millennium cal BCE with evidence for bones of small cetaceans (Dupont et al. 2009). Another site with bone preservation is the megalithic grave La Planche à Puare on the Atlantic island of Île d'Yeu. Excavations there in 1883 revealed three sperm whale teeth in one of the side chambers of the passage grave (Baudouin 1907).

Recent recordings with a HandySCAN 700 red-light laser scanner in the megalithic grave Mané Rutual within the research program Corpus des signes gravés néolithiques PCR revealed a sperm whale and possible whaling scenery. Mané Rutual is a 15-m-long passage grave with reused standing stones as coverage and partly orthostats (i.e., large stones with a slab-like shape that constitute part of the walls of the larger structure). Sperm whales are engraved on two of these orthostats. The laser scan of the orthostat inside the grave showed a small boat in front of the head with two crew strokes reminiscent of paintings of historical sperm whale hunts (Figure 4.2).

Stone Age Whalers at the Chilean Atacama Desert Coast

One of the most arid and extreme terrestrial environments on Earth is the arid coast of northern Chile (Rebolledo et al. 2015). At the same time, this area is among the most environmentally diverse marine ecosystems worldwide. It thrives due to the cold Humboldt Current and resulting upwellings, creating an environment that sustains a wide variety of marine life (Thiel et al. 2007). This has led to ongoing use of the region's marine resources for the past 13,000 years (Olguín et al. 2015).

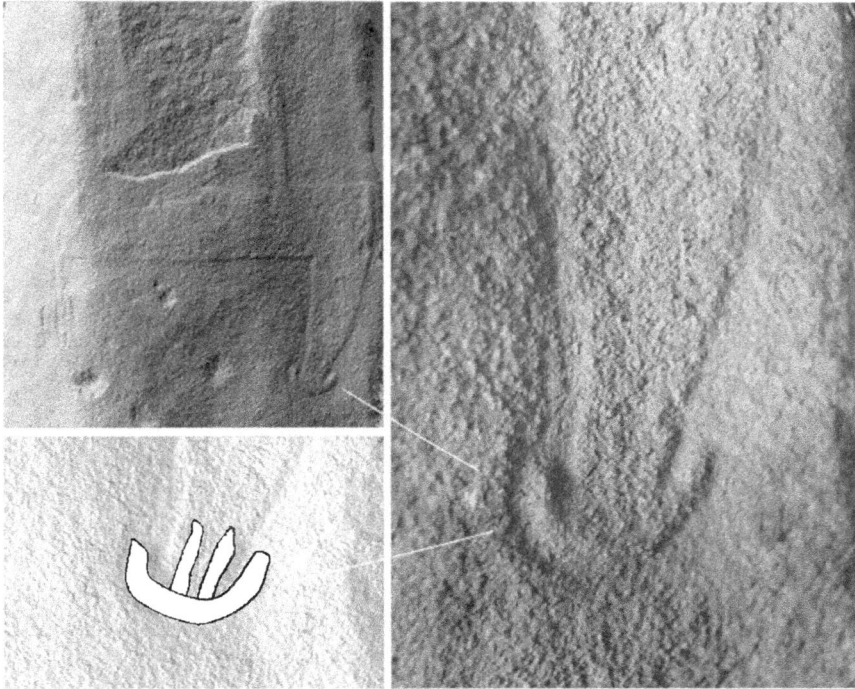

Figure 4.2. Passage Grave Mané Rutual in Locmariaquer, Brittany. The orthostat in the middle of the grave is a reused standing stone with an engraving of a sperm whale. The laser scan revealed a small boat in front of the head with two crew. Visualized with the software MeshLab.

The hyperarid climatic conditions of the area provide optimal circumstances for the preservation of organic materials. Consequently, it is the sole region in the world where it is possible to directly link marine hunting rock art with the tools and even the remains of boats. Hundreds of rock paintings in red ocher depict marine fauna, boats, and hunting sceneries in a naturalistic manner, providing insights into hunting technologies and strategies of these coastal societies.

In recent years, the number of discoveries of the so-called El Médano rock art tradition, named after the eponymous site El Médano, has significantly increased (Ballester 2018). Most of these rock paintings are concentrated in a desert setting, located within narrow and challenging-to-access ravines situated several kilometers from the shoreline. Only a few have been identified in rock shelters near the coast. These sites include not only El Médano but also Las Cañas, Quebrada de la Plata, Izuñja, and Botija.

The predominant subject of these paintings consists of various species of fish and sea mammals. In some cases it is possible to distinguish specific species, including well-recognizable depictions of whales, sea lions, turtles, sharks, squid, swordfish, and marlin (Ballester and Gallardo 2016; Berenguer 2009; Gallardo et al. 2012; Niemeyer 2010). Among these depictions of marine fauna, marine hunting scenes are the most commonly identified motif (Ballester 2018).

Of particular significance are the whaling scenes, featuring the hunting of large whale species using harpoons, typically from small boats manned by one to four crew members. These hunting scenes involve a single boat, whether fitted with one or multiple harpoons, shown through the ropes (Figure 4.3). The depictions of whales are often exaggerated in size, while the boats, featuring either one or several oars or a small mast, are recurring motifs. Additionally, the artwork may include representations of camels, geometric patterns, oars, and baskets. The earliest AMS radiocarbon dates for this art are from the first half of the seventh millennium calibrated BCE (Castelleti 2007), but the rock art tradition here seems to have survived for several thousands of years.

The archaeological artifacts and preserved organic remains attest to maritime technologies and sophisticated hunting strategies. Within the artifact assemblages are fishhooks made from bone, cactus spines, shells and copper, cotton ropes, nets, net weights, oars, harpoons, and remains of boats (e.g., Ballester and Clarot 2014; Castelleti 2007; Mostny 1964; Salazar et al. 2015). Entire harpoons are preserved from residential contexts and, later, in graves (Ballester 2018). The harpoons were up to 3-m-long wooden shafts with composite harpoon heads made of lithic points, copper barbs, and different raw materials such as wood stems, cactus spine barbs, cotton strings, and vegetal resin with a pigment coating. Attached to these harpoons were up to 70-m-long ropes made of sea lion skin and suitable for hunting large game. Also made of sea lion skin are the remains of skin boats, such as from the El Trocadero cemetery, where a float fragment with spine seam was found (Ballester et al. 2015).

Whaling and the Elk Boat People of Fennoscandia

Sea mammals are a recurring motif in Stone Age rock art across northern Fennoscandia. Examples of such depictions can be found at sites like Bogge 2, Skelvejen, and Hammar IV in central Norway. For instance, on the Hammar IV panel, there are at least seven porpoises depicted in association with small boats. Another hunting scene featuring a smaller boat, a harpoonist, and a seal is documented from Bergbukten 1, Alta, in northern Norway. The southernmost example of this phenomenon is found at Tumlehed, north of Gothenburg in southern Scandinavia, where evidence exists for a red painting featuring a

Figure 4.3. El Médano whaling rock art, Atacama Coast, Chile. Photo reproduced with the kind permission of Francisco Gallardo.

porpoise and a seal alongside three boats (Schulz Paulsson et al. 2019; see Figure 4.4). Sea mammals on Scandinavian rock art are mostly depicted together with elk-head stemmed boats, which we find as engravings and paintings in the coastal areas of northeast Norway and northern Sweden, with a few found in southern Sweden, Finland, and associated areas of northwest Russia, but with a clear eastern emphasis (Gjerde 2010: Figure 283).

The elk-head stemmed boats serve as a link between the coastal societies of the Late Stone Age in the subpolar and adjacent regions during the sixth and third millennia cal BCE. These societies had a strong reliance on marine resources and the hunting of sea mammals, with the most substantial evidence stemming from the Kola Peninsula in Russia. Clear whaling sceneries involving several actors whaling from boats with an elk-headed stem are known from Vyg, Kanozero, and Lake Onega.

At the estuary of the River Vyg, near the White Sea in northwestern Russia, lies one of the most remarkable collections of Stone Age rock art in northern Europe, with over 2,300 documented figures (Gjerde 2010, 2013, 2019; Janik 2010, 2017, 2018, 2019, 2022; Savvateev 1970). What sets the Vyg rock art apart is its abundant representation of scenes depicting whale hunting. These highly detailed depictions enable us not only to make observations about the technology of whaling but also to glean insights into the organization of this hunt and the development of hunting strategies in this region over time. In the Vyg region are more than 60 depictions of whaling scenes as well as a concentration in Kanozero and, intriguingly, even more inland, one a scene of whaling around Lake Onega (Gjerde 2013). These engravings predominantly illustrate the hunt of belugas (*Delphinapterus leucas*), also referred to as white whales. Such interpretations are grounded in considerations of morphological and typological characteristics as well as the notable presence of these marine mammals within the White Sea ecosystem today (Janik 2022).

The dating of this rock art falls within the range of 5300–2000 cal BCE, although it is important to note that the chronology is primarily reliant on shoreline data and remains a subject of ongoing debate (Gjerde 2013). The shoreline chronology involves the reconstruction of the depicted scenes in proximity to the water, incorporating C-14 dating from associated contexts. While the refinement of the chronology requires additional data, it nevertheless provides valuable insights into the evolution of whale hunting practices over time. Gjerde (2010) has proposed three distinct phases.

In the earliest phase, Phase 1, dating to approximately 5300–4200 cal BCE, the rock art depicts whaling scenarios that feature a singular boat with a crew of up to three individuals. Most of the boats are illustrated with elk-headed stem ornamentation. Significantly, these depictions consistently show the presence of a lone harpoon line, and the hunting activities encompass either solitary whale pursuits or, in certain instances, the pursuit of multiple whales, with a maximum recorded count of three.

The harpoonist assumes a position at the front of the boat. The whales depicted in this phase often appear somewhat exaggerated in size, possibly re-sembling totems or suggesting the hunting of other whale species. The harpoon rope is frequently seen trailing from the tail of the whale, making it unclear whether the depiction represents the actual killing of the whale or the process of hauling back the hunted and deceased animal. The alignment of this phase with the updated chronology in Alta, based on recent shoreline radiocarbon data, places it within a time interval of 5220–4240 cal BCE. As in the Vyg region, the elk boats in this phase are relatively modest in size, typically featuring two to five crew members.

In Phase 2, dated between 3700 and 3400 cal BCE, notable developments emerge. Elk-head boats become larger, accommodating as many as 12 crew members. This phase also marks the first depictions of collective whale hunt-

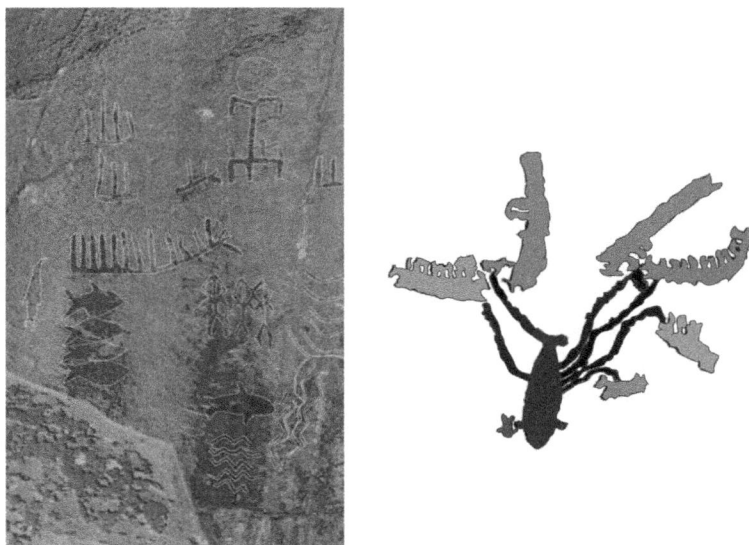

Figure 4.4. Fennoscandia whaling rock art. *Left*: Detail with elk-headed boats and marine fauna of the Tumlehed Stone Age rock painting panel, north of Gothenburg, Sweden. *Right*: Detail whaling scenery at Zalavruga 8 (after Savvateev 1970, modified in different greyscales).

ing, as up to three boats are shown surrounding and harpooning the whales. Among these depictions, the Kanozero panel stands out as the most sophisticated example. Twelve whalers stand in a boat with their legs bent. Each of them holds either a harpoon or a club in their hands. It is possible that there are multiple harpoonists, or the clubs were used to weaken or kill the animals. The crew is depicted with intricate detail. The crew members are portrayed with a topknot-like hairstyle. The first man is the harpoonist and the last one, the helmsman, is depicted with a type of hat or other hair adornment, along with a weapon or erect penis. In Alta, larger elk-head boats with a bigger crew are also depicted in Phase 2 within a 4230–2920 cal BCE range.

In Vyg Phase 3, spanning from approximately 3400 to 2500 cal BCE, hunting strategies become even more sophisticated. Up to six boats encircle and pursue a whale, with each boat deploying a harpoon line in an attempt to strike the animal. Interestingly, the boats do not appear noticeably larger than in Phase 2, provided we can accurately position these larger boats toward the end of the second phase. The depictions include up to 12 crew members (Figure 4.4).

Real Hunting, Mythological Narratives or Observations, and the Question of Scavenging Stranded Whales

The presented case studies affirm that whaling was integral to the subsistence of coastal communities during the middle Holocene. These whaling societies exhibited advanced hunting strategies, challenging the notion that whale bones in Stone Age western European archaeological assemblages necessarily resulted from scavenging stranded animals (e.g., Charpentier et al. 2022; Pétillon 2016; Schuhmacher et al. 2013).

Most of all, the Chilean evidence characterized by the remarkable preservation of organic remains indicates that rock art represents actual hunting activities instead of mythological narratives. The desert's arid climate has effectively conserved hunting tools such as entire harpoons, boat components, ropes, and floats. The Fennoscandian rock art showcases a degree of detail, incorporating technical nuances, such as the representation of harpoons and floats. This level of detail also suggests an emphasis on practical hunting techniques as opposed to mythological storytelling.

There is evidence from both regions indicating that rock art was produced inland rather than directly on the coast. Although paintings can be found in a few rock shelters near the shore at the Atacama Desert, they are far more prevalent in ravines located several kilometers from the ocean (Ballester 2018: 145). These ravines are often narrow and challenging to access, and the artists depicted maritime activities in a desert environment. These two distinct settings were also connected by the technology used to hunt whales, as the harpoons and rafts were crafted from materials sourced from the inland desert, such as lithics, red ocher, wood, cotton, resin, cactus, and camelid bone. This suggests that the coast was primarily used as hunting grounds, while the inland areas served as settlements or, at the very least, seasonal camps. Similar patterns are observed in Fennoscandia, particularly on the Kola Peninsula. Most whaling scenes are not depicted directly along the White Sea but are predominantly clustered in four areas surrounding the estuary of the River Vyg. The Vyg links Lake Onega to the White Sea. The presence of a lone whaling scene at Lake Onega, situated 300 km inland from the whaling rock art along the Vyg River, indicates a seasonal hunting pattern (Gjerde 2010, 2013). During the summer months whaling activities occurred in the shallow waters, possibly accompanied by summer camps, while in the winter, settlements were established farther inland. Rock art in this context serves to convey the narratives of these real hunting seasons.

Various cetacean species require different hunting strategies, and it is rock art that can provide important insights into the manner of this exploitation and the maritime technologies of these communities (Erlandson 2001; Mulville

2002). In both regions, whales of different sizes are depicted, although it is not always clear if these are animistic oversized paintings or larger cetacean species.

For the Atacama Desert, it is difficult to define the species of the cetaceans. Today there are large whale species known, such as fin whale (*Balaenoptera physalus*), southeast Pacific blue whales (*Balaenoptera musculus*), sperm whale (*Physeter macrocephalus*), as well as dolphins such as the dusky dolphin (*Lagenorhynchus obscurus*) and the common dolphin (*Delphinus* spp.). The marine animal bone assemblages in the prehistoric settlements include more dolphins (*Delphinidae*), sharks (such as *Notorynchus cepedianus, Galeorhinus galeus, Isurus oxyrinchus*), and smaller cetaceans (e.g., *Phocoena spinipinnis*) (Béarez et al. 2016; Mostny 1942; Rebolledo et al. 2015). There are also examples of seals that are clearly painted in an exaggerated style compared to the boats and the humans, thus use as a starting point at least a part of oversized paintings in the Atacama Desert.

For Fennoscandia, it is possible to define the species more precisely. From the morphological features of the animals, it seems that the rock art portrays the hunting of beluga whales (*Delphinapterus leucas*) (Gjerde 2010, 2013), which continue to be a prominent part of the current ecosystem in the White Sea. Belugas are the most common species depicted (Boltunov and Belikov 2002: 150), while species such as bowhead whales, humpback whales, rorquals, northern bottlenose whales, and orcas are seldom depicted.

The presence of belugas in the White Sea exhibits a distinct seasonality, closely tied to the migratory patterns of their primary prey species, including Navaga (*Eleginus nawaga*), White Sea herring (*Clupea pallasii marisalbi*), and smelt (*Osmerus eperlanus*) (Janik 2022; Svetochev and Svetocheva 2013).

Belugas still tend to aggregate in areas proximate to the rock art locations. During the summer months these cetaceans migrate into the shallower coastal waters for mating, breeding, and parturition, rendering these locales conducive for targeted hunting activities. Adult individual beluga whale lengths vary between 3.5 m and 5.5 m (O'Corry-Crowe 2009; O'Corry-Crowe et al. 2000). During migrations from their feeding grounds in the north, beluga whales occasionally form extensive herds consisting of several hundred individuals. The shallow coastal areas offered optimal conditions for prehistoric hunting. Small whales and dolphins normally occur in flocks and could be best hunted either by a group of boats surrounding them and harpooning them or by chasing them into a bay and killing them in shallow water with the help of actors from the land. For large whale species there are other hunting strategies required involving surrounding the animal with several boats on the open sea and harpooning it with harpoons with floats on them.

There are numerous ethnohistorical examples for this kind of whaling, including among the whaling people along the west coast of Vancouver Island and the Cane Flattery (Arima and Hoover 2011). These communities hunted various large whale species—primarily gray, humpback, and occasionally sperm whales—using multiple canoes with a large amount of equipment and involving complex ritual preparations. Of significant importance was the principal whaling canoe, for a crew of 6–12 individuals, including the harpooner, the steersman, and at least 6 paddlers. This main vessel was followed by additional whaling canoes led by younger kinsmen of high social rank. A swift sailing canoe was also on standby to relay news to the village upon harpooning the whale. The initial harpooning of the whale took place from the primary canoe using ropes up to 110 m long and made from three-stranded cedar or split spruce rope, along with floats crafted from entire skins of harbor seals (Arima and Hoover 2011: 61). Subsequently, other canoes would approach to plant their harpoons into the weakened animal.

This kind of whaling is depicted in the later phases of the Fennoscandian rock art like in New Zalavruga 8 (Gjerde 2010: Figure 216). Six boats of different size and with crews between 2 and 12 people surround one large whale and kill it in a common harpooning effort. This depiction appears to illustrate the hunting of one of the larger whale species, which are less frequently seen in the White Sea region, rather than the oversized depiction of a beluga.

A similar hunting method for sperm whales continues to be practiced on the island of Lamalera in Indonesia. Villages there have upheld traditional hunting practices to this day. This technique involves several boats surrounding the whale and a harpooner who jumps from a platform at the front of the boat onto the whale, thrusting the harpoon directly into the animal (e.g., Alvard and Gillespie 2015; Lundberg 2003).

Regarding the Breton sperm whale engravings, it is challenging to prove the hunting and consumption of the depicted sperm whales in this region. Sparse bone preservation in the area results in very few whale bones, and there is an absence of human bones in the region.

To define a marine mammal diet, we would need to calculate the δ^{13}C or δ^{34}S values on the human bones. For the pre-megalithic Mesolithic contexts in Brittany, there are radiocarbon determinations as well as stable isotope results available for the island of Téviec and six from the island of Hoëdic (Schulting and Richards 2001: Table 2, cf. Table 3.2). The radiocarbon results of the different individuals from Téviec hint to a time interval from 5736 to 4372 cal BCE with δ^{13}C values from -17.0‰ to -14.1‰. The radiocarbon determinations from Hoëdic lie within a time interval of 6211–3714 cal BCE with δ^{13}C values from -14.2‰ to -12.9‰. The results from both sites show that a heavy marine-based diet is customary, including the consumption of sea mammals (Schulz Paulsson 2017: 32).

Clear possible hunting tools are so far not recorded, but also here, we have to take into consideration that prehistoric tools for hunting sea mammals are often fabricated out of bone or antler (e.g., Ballester 2018; Pétillon 2016), and they might—like the human and the faunal bones—not be preserved in the acidic soils of the Morbihan.

However, we cannot entirely exclude the slaughter of stranded animals. Stranded animals have consistently provided an economic surplus throughout history (e.g., Castelleti 2020; Gusinde 1931; Gutierrez et al. 2001; Pétillon 2016; Rodrigues et al. 2016; Schnall 1992). Sperm whales are among the whale species that frequently strand alive on beaches or in shallow waters, unable to return to deeper waters with the rising tide. The stranding of sperm whales on North Sea shores is linked, for instance, to temperature anomalies and a warmer climate. These factors lead to altered migration patterns of their main food source, the squid (Pierce et al. 2007). Instances where sperm whales become stranded today in Brittany are described as extremely rare, which is also the case when the animals are washed ashore dead. These animals were undoubtedly challenging to utilize and were likely harvested, for instance, for their blubber.

As mentioned earlier, the sperm whale engravings on the standing stones depict living animals in the water. The blow is depicted, and the fluke is engraved in a three-dimensional manner, resembling their appearance when whales are swimming close to the surface (Cassen and Vaquero 2000). Together with the recently discovered carvings of a boat close to the jaw of these whales, it becomes evident that these societies must have at least had encounters with sperm whales in the open sea and a developed boat technology suitable for open-sea navigation.

Conclusion

This chapter demonstrates that whaling was an important economic factor for coastal societies in the Stone Age. Just as rock art around the world has depicted communal hunts of terrestrial game, experiences of communal hunting expressed in rock art provide us with insights into the organization of whaling practice. It is evident that it was also possible to hunt larger whale species, not just those that were stranded or driven into shallow waters in bays. The depiction of sperm whales on standing stones and in megalithic tombs in the Gulf of the Morbihan region and in northern Brittany, particularly when correlated with the representation of small boats, indicates scenes that lean toward whaling activities rather than the mere depiction of observing or harvesting stranded whales.

Acknowledgments

I would like to thank the editors, Colin Grier, Mikael Fauvelle, and Alberto García-Piquer, for inviting me to participate in this volume. Furthermore, I would like to express my gratitude to Francisco Gallardo, Center for Intercultural and Indigenous Research, Universidad Católica de Chile, for providing the photo of the El Médano rock art.

This research is funded by the European Union and the European Research Council Executive Agency (ERC, NEOSEA, 949424). Views and opinions expressed are, however, those of the author(s) only and do not necessarily reflect those of the European Union or the European Research Council Executive Agency. Neither the European Union nor the granting authority can be held responsible for them.

References Cited

Alvard, Michael S., and Allen Gillespie. 2015. Good Lamalera Whale Hunters Accrue Reproductive Benefits. *Socioeconomic Aspects of Human Behavioral Ecology* 23: 225–247.

Arima, Eugene, and Alan Hoover. 2011. *The Whaling People of the West Coast of Vancouver Island and Cape Flattery.* Royal British Columbia Museum, Vancouver.

Ballester, Benjamín. 2018. El Médano Rock Art Style: Izcuña Paintings and the Marine Hunter-Gatherers of the Atacama Desert. *Antiquity* 92(361): 132–148.

Ballester, Benjamín, and Alejandro Clarot. 2014. *La gente de los Túmulos de Tierra: Estudio, Conservación y Difusión de Colecciones Arqueológicas de la Comuna de Mejillones.* Marmot Impresores, Santiago.

Ballester, Benjamín, and Francisco Gallardo. 2016. Painting a Lost World: The Red Rock Art of El Médano. *Current World Archaeology* 77: 36–38.

Ballester, Benjamín, Francisco Gallardo, and Patricio Aguilera. 2015. Representaciones que navegan más allá de sus aguas: Una pintura estilo El Médano a más de 250km de su sitio homónimo. *Boletín de la Sociedad Chilena de Arqueología* 45: 81–93.

Baudouin, Marcel. 1913. Découverte et fouille d'un mégalithe funéraire aux Landes, à L'Ile d'Yeu (Vendée). *Bulletins et Mémoires de la Société d'anthropologie de Paris,* VI Série, 4(2): 195–208.

Béarez, Philippe, Felipe Fuentes-Mucherl, Sandra Rebolledo, D. Diego Salazar, and Laura Olguín. 2016. Billfish Foraging along the Northern Coast of Chile During the Middle Holocene (7400–5900 cal BP). *Journal of Anthropological Archaeology* 41: 185–195.

Berenguer, José R. 2009. Las pinturas de El Médano, norte de Chile: 25 años después de Mostny y Niemeyer. *Boletín del Museo Chileno de Arte Precolombino* 14(2): 57–95.

Boethius, Adam. 2018. Fishing for Ways to Thrive: Integrating Zooarchaeology to Understand Subsistence Strategies and Their Implications among Early and Middle Mesolithic Southern Scandinavian Foragers. PhD dissertation, Department of Archaeology and Ancient History, Lund University, Lund.

Boltunov, Andrei N., and Stanislav E. Belikov. 2002. Belugas (*Delphinapterus leucas*) of the Barents, Kara and Laptev seas. In *Belugas in the North Atlantic and the Russian Arctic*, edited by Mads Peter Heide-Jørgensen and Øystein Wiig, pp. 149–169. North Atlantic Marine Mammal Commission, Tromsø.

Buckley, Michael, Sheena Fraser, Jeremy S. Herman, Nigel D. Melton, Jacqui A. Mulville, and A. H. Pálsdóttir. 2014. Species Identification of Archaeological Marine Mammals Using Collagen Fingerprinting. *Journal of Archaeological Science* 41: 631–641.

Cassen, Serge, Christine Boujot, Valentin Grimaud, Olivier Célo, Cyrille Chaigneau, Christian Obeltz, and Emmanuelle Vigier. 2021. *Carnac. Récit pour un imagier*. Laboratoire de recherche archéologie et architectures, Nantes.

Cassen, Serge, and Valentin Grimaud. 2020a. *La Clef de la mer: Une étude des représentations gravées sur la Pierre de Saint-Samson (Côtes-d'Armor)*. Lithogénies 1—LARA/ Université de Nantes, Nantes.

Cassen, Serge, and Valentin Grimaud. 2020b. Nouvelles gravures néolithiques dans la tombe à couloir de Cruguellic (Ploemeur, Morbihan). *Bulletin Société d'Archéologie et d'Histoire du Pays de Lorient (2016–2019)* 45: 51–68.

Cassen, Serge, and Jacobo Vaquero. 2000. La Forme d'une Chose. In *Eléments d'Architecture*, edited by Serge Cassen, Christine Boujot, and Jacobo Vaquero, pp. 611–656. Association des Publications chauvinoises, Chauvigny.

Castelleti, José. 2007. Patrón de asentamiento y uso de recursos a través de la secuencia ocupacional prehispánica en la costa de Taltal. Master's thesis, Universidad Católica del Norte, Antofagasta.

Castelleti, José. 2020. Whale Strandings or Hunting and the Making of El Médano Style Paintings in the Atacama Desert Coast in Chile. *Cuadernos de Arte Prehistórico* 1: 215–255.

Charpentier, Anne, Ana Rodrigues, Claire Houmard, Alexandre Lefebvre, Krista McGrath, Camilla Speller, Laura van der Sluis, Antoine Zazzo, and Jean-Marc Pétillon. 2022. What's in a Whale Bone? Combining New Analytical Methods, Ecology and History to Shed Light on Ancient Human–Whale Interactions. *Quaternary Science Reviews* 284: 107470.

Clark, Grahame A. 1947. Whales as an Economic Factor in Prehistoric Europe. *Antiquity* 21: 84–104.

Colonese, André C., Marcello A. Mannino, Daniella E. Bar-Yosef Mayer, Darren Andrew Fa, Clive J. Finlayson, David Lubell, and Mary C. Stiner. 2011. Marine Mollusc Exploitation in Mediterranean Prehistory: An Overview. *Quaternary International* 239 (1–2): 86–113.

Dupont, Catherine, Anne Tresset, Nathalie Desse-Berset, Yves Gruet, Grégor Marchand, and Rick Schulting. 2009. Harvesting the Seashores in the Late Mesolithic of Northwestern Europe: A View from Brittany. *Journal of World Prehistory* 22: 93–111.

Erlandson, Jon M. 2001. The Archaeology of Aquatic Adaptations: Paradigms for a New Millennium. *Journal of Archaeological Research* 9(4): 287–350.

Fenton, Alexander. 1997. *The Northern Isles: Orkney and Shetland*. Tuckwell Press, East Linton.

Gallardo, Francisco, Gloria Cabello, Gonzalo Pimentel, Marcela Sepúlveda, and Luís Cornejo. 2012. Flujos de información visual, interacción social y pinturas rupestres en el desierto de Atacama (norte de Chile). *Estudios Atacameños* 43: 35–52.

Gjerde, Jan Magne. 2010. Rock Art and Landscapes: Studies of Stone Age Rock Art from Northern Fennoscandia. PhD dissertation, Department of Archaeology and Social Anthropology, University of Tromsø, Tromsø.

Gjerde, Jan Magne. 2013. Stone Age Rock Art and Beluga Landscapes at River Vyg, North-Western Russia. *Fennoscandia Archaeologica* 30: 37–54.

Gjerde, Jan Magne. 2019. An Overview of Stone Age Rock Art in Northernmost Europe—What, Where and When? In *Rock Art of the White Sea*, edited by Liliana Janik, pp. 205–209. Ulsan Petroglyph Museum, Ulsan.

Gusinde, Martin. 1931. *Die Feuerlandindianer: Ergebnisse meiner vier Forschungsreisen in den Jahren 1918 bis 1921 unternommen im Auftrage des Ministerio de instruccion publica de Chile. Band Die Halakwulup. Vom Leben und Denken der Wassernormaden in Westpatagonien.* Verlag der Internationalen Zeitschrift "Anthropos," Mödling bei Wien, 294–295.

Gutierrez, Manuel, Claude Guérin, Maria Léna, and Maria Piedade da Jesus. 2001. Human Exploitation of a Large Stranded Whale in the Lower Palaeolithic Site of Dungo V at Baia Farta (Benguela, Angola). *Comptes Rendus de l'Académie des Sciences—Series IIA—Earth and Planetary Science* 332(5): 357–362.

Herman, Jeremy S., and Keith M. Dobney. 2004. Evidence for an Anglo-Saxon Dolphin Fishery in the North Sea. *European Research on Cetaceans* 15: 161–165.

Janik, Liliana. 2010. The Development and Periodisation of White Sea Rock Art Carvings. *Acta Archaeologica* 81: 83–94.

Janik, Liliana. 2017. Rock Art as an Independent Evidence of Prehistoric Marine Hunting. In *Whale on the Rock*, edited by Sang-mog Lee, pp. 169–182. Ulsan Petroglyph Museum, Ulsan.

Janik, Liliana. 2018. The Unique and the Common: The Rock Art of the White Sea. In *Rock Art of the White Sea*, edited by Liliana Janik, pp. 52–68. Ulsan Petroglyph Museum, Ulsan.

Janik, Liliana. 2019. The Ontology of Praxis: Hard Memory and the Rock Art of the White Sea. *Time and Mind* 12(3): 207–219.

Janik, Liliana. 2022. Visual Narratives and the Depiction of Whaling in North European Rock Art: The Case of the White Sea. *Les nouvelles de l'archéologie* 166: 51–63.

Kristjánsson, Lúðvík. 1986. *Íslenskir sjávarhættir.* Vol. 5. Menningarsjóður, Reykjavík.

Lundberg, Anita. 2003. Time Travels in Whaling Boats. *Journal of Social Archaeology* 3(3): 312–333.

Mostny, Greta. 1942. Informe preliminar sobre las excavaciones efectuadas en la costa chilena entre Pisagua y Coquimbo del 8 de octubre de 1941 al 15 de marzo de 1942. *Boletín del Museo Nacional de Historia Natural* 20: 97–102.

Mostny, Greta. 1964. *Arqueología de Taltal: Epistolario de Augusto Capdeville con Max Uhle y otros.* Fondo Histórico y Bibliográfico José Toribio Medina, Santiago.

Mulville, Jacqui. 2002. The Role of Cetacea in Prehistoric and Historic Atlantic Scotland. *International Journal of Osteoarchaeology* 12(1): 34–48.

Niemeyer, Hans. 2010. *Crónica de un descubrimiento: Las pinturas rupestres de El Médano, Taltal.* Museo Chileno de Arte Precolombino, Santiago.

O'Corry-Crowe, Greg. 2009. Beluga Whale: Delphinapterus leucas. In *Encyclopedia of Marine Mammals,* 2nd ed., edited by William F. Perrin, Bernd Würsig, and J.G.M. Thewissen, pp. 108–112. Academic Press.

O'Corry-Crowe, Greg, Robert Suydam, Lori Quakenbush, Thomas G. Smith, Christian Lydersen, Kit M. Kovacs, Jack Orr, Lois Harwood, Dennis Litovka, and Tatiana Ferrer. 2000. Group Structure and Kinship in Beluga Whale Societies. *Scientific Reports* 10: 11462.

Olguín, Laura, Victoria Castro, Pilar Castro, Isaac Peña-Villalobos, Jimena Ruz, and Borís Santander. 2015. Exploitation of Faunal Resources by Marine Hunter-Gatherer Groups During the Middle Holocene at the Copaca 1 Site, Atacama Desert Coast. *Quaternary International* 373: 4–16.

Pétillon, Jean-Marc. 2016. Life on the Shore of the Bay of Biscay in the Late Upper Paleolithic: Towards a New Paradigm. In *Archaeology of Maritime Hunter-Gatherers: From Settlement Function to the Organization of the Coastal Zone,* edited by Catherine Dupont and Grégor Marchand, pp. 23–34. Séances de la Société Préhistorique Française 6. Société préhistorique française, Paris.

Pierce, Graham J., Maria Begoña Santos, Chris Smeenk, Anatoly A. Saveliev, and Alain François Zuur. 2007. Historical Trends in the Incidence of Strandings of Sperm Whales (*Physeter macrocephalus*) on North Sea Coasts: An Association with Positive Temperature Anomalies. *Fisheries Research* 87(2–3): 219–228.

Rebolledo, Sandra, Philippe Béarez, Diego Salazar, and Felipe Fuentes. 2015. Maritime Fishing during the Middle Holocene in the Hyperarid Coast of the Atacama Desert. *Quaternary International* 391: 3–11.

Rodrigues, Ana S. L., Liora Kolska Horwitz, Sophie Monsarrat, and Anne Charpentier. 2016. Ancient Whale Exploitation in the Mediterranean: Species Matters. *Antiquity* 90(352): 928–938.

Salazar, Diego, Valentina Figueroa, Pedro Andrade, Hernán Salinas, Laura Olguín, Ximena Power, Sandra Rebolledo, Sonia Parra, Héctor Orellana, and Josefina Urrea. 2015. Cronología y organización económica de las poblaciones arcaicas de la costa de Taltal. *Estudios Atacameños* 50: 7–46.

Savvateev, Yuri A. 1970. *Zalavruga, Petroglify.* Nauka, Leningrad.

Schnall, Uwe. 1992. Der Kampf um die "Gabe Gottes": Auseinandersetzungen über gestrandete Wale in Nordeuropa zur Wikingerzeit. *Deutsches Schiffsarchiv* 15: 209–222.

Schuhmacher, Thomas X., Arun Banerjee, Willi Dindorf, Chaturvedula Sastri, and Thierry Sauvage. 2013. The Use of Sperm Whale Ivory in Chalcolithic Portugal. *Trabajos de Prehistoria* 70(1): 185–203.

Schulting, Rick J., and Michael P. Richards. 2001. Dating Women and Becoming Farmers: New Palaeodietary and AMS Dating Evidence from the Breton Mesolithic Cemeteries of Téviec and Hoëdic. *Journal of Anthropological Archaeology* 20(3): 314–344.

Schulz Paulsson, Bettina. 2017. *Time and Stone: The Emergence and Development of Megaliths and Megalithic Societies in Europe.* Archaeopress, Oxford.

Schulz Paulsson, Bettina. 2019. Radiocarbon Dates and Bayesian Modelling Support Maritime Diffusion Model for Megaliths in Europe. *Proceedings of the National Academy of Sciences of the United States of America (PNAS)* 116(9): 3460–3465.

Schulz Paulsson, Bettina. 2025. Transmission symbolique et rencontres transculturelles maritimes dans l'Europe mégalithique (4700–2500 Cal BC). In Les Gravures Rupuestres Protohistoriques en Eurasie. Actes du Colloques International du mardi 7 au 9 décembre 2021 á L'Institut de Paléóntologie Humaine, Paris. Sous le haut patronage de S.A.S Le Prince Souverain Albert II de Monaco, 471, edited by H. Lumley and A. Echassoux.

Schulz Paulsson, Bettina, Christian Isendahl, and Fredrik Frykman Markurth. 2019. Elk Heads at Sea: Maritime Hunters and Long-Distance Boat Journeys in Late Stone Age Fennoscandia. *Oxford Journal of Archaeology* 38(4): 1–24.

Shee Twohig, Elizabeth. 1981. *The Megalithic Art of Western Europe*. Clarendon Press, Oxford.

Sinding, Mikkel-Holger S., M. Thomas, P. Gilbert, Bjarne Grønnow, Hans Christian Gulløv, Peter A. Toft, and Andrew D. Foote. 2012. Minimally Destructive DNA Extraction from Archaeological Artefacts Made from Whale Baleen. *Journal of Archaeological Science* 39(12): 3750–3753.

Storå, Jan, and Lembi Lõugas. 2005. Human Exploitation and History of Seals in the Baltic during the Late Holocene. In *The Exploitation and Cultural Importance of Sea Mammals,* edited by Gregory G. Monks, pp. 95–106. Oxbow, Oxford.

Stringer, Christopher B., Clive J. Finlayson, Nick E. Barton, Yolanda Fernández-Jalvo, Isabel Cáceres, Richard Sabin, E. Rhodes, et al. 2008. Neanderthal Exploitation of Marine Mammals in Gibraltar. *Proceedings of the National Academy of Science of the United States of America (PNAS)* 105(38): 14319–14324.

Svetochev, Vladislav N., and O. N. Svetocheva. 2013. Summer Migration Activity of the Beluga Whale *Delphinapterus leucas* in Dvina Bay, the White Sea. *Doklady Biological Science* 448: 17–21.

Szabo, Vicki E. 2008. *Monstrous Fishes and the Mead-Dark Sea*. Brill, Leiden.

Thiel, Martin, Erasmo C. Macaya, Enzo Acuña, Wolf E. Arntz, Horacio Bastias, Katherina Brokordt, Patricio A. Camus, et al. 2007. The Humboldt Current System of Northern and Central Chile: Oceanographic Processes, Ecological Interactions and Socioeconomic Feedback. *Oceanography and Marine Biology: An Annual Review* 45: 195–344.

Whittle, Alasdair. 2000. "Very Like a Whale": Menhirs, Motifs and Myths in the Mesolithic–Neolithic Transition of Northwest Europe. *Cambridge Archaeological Journal* 10(2): 243–259.

5

Seascapes and Society on the Forgotten Peninsula

The Watercraft and Conceptual Geography of Baja California, Mexico

MATTHEW R. DES LAURIERS AND
CLAUDIA GARCÍA-DES LAURIERS

Baja California is a unique landscape of over 3,000 km of coastline covering a diversity of arid temperate zones to the northern edge of the tropics adorned by rugged mountain ranges, effusive desert flora, and fecund seascapes. Early visions presented this original "California" as a literal island, separate from the mass of the North American continent. This mythical geography (Mathes 1989) of literary creation still frames the human experience of this landscape since no human group has ever set foot on the Baja California Peninsula without having knowledge of the sea. It is nearly impossible to find locations in Antigua California that are not within sight of, or significantly influenced by, the pelagic expanses that circumscribe it.

Two radically different "seas" shove insistently upon the narrow peninsula (Figure 5.1). We are certainly not the first to comment on the unique geography of Baja California, but it is worth highlighting the sheer diversity of water temperatures, resident species, characteristics of prevailing climate, varying timing for seasonal transitions, weather patterns, wind directions, and tidal ranges that exist with a mere 50 km separating wide extremes in some places. A greater contrast in marine conditions may not exist in such proximity anywhere in the world, meaning that for the Indigenous populations, the experience of the eastern sea and western ocean would have been markedly different, especially for those groups in the northern two-thirds of the peninsula.

We propose that the spaces, which provide the context for both thought and action in pre-sixteenth-century Baja California, were less the "landforms" that a Western cartographer would map (Edney 2019; see, e.g., Figure 5.2) and

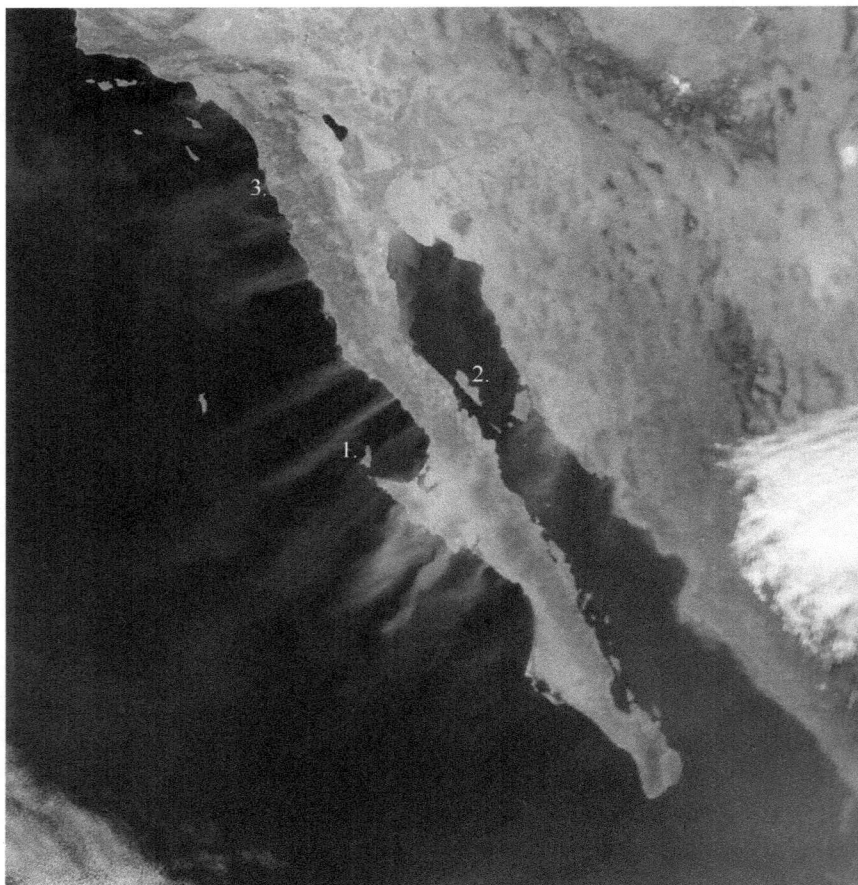

Figure 5.1. The Baja California Peninsula with points mentioned in the text: (1) Isla Cedros; (2) Isla Angel de la Guarda; and (3) Las Islas Coronados. Image courtesy the SeaWiFS Project, http://seawifs.gsfc.nasa.gov/SEAWIFS.html, NASA/Goddard Space Flight Center, and ORBIMAGE, https://eoimages.gsfc.nasa.gov/images/imagerecords/2000/2184/BajaDust_2002041_lrg.jpg.

instead a complex chain of interfaces between Homer's "wine-dark sea" (see Maxwell-Stuart 1981) itself and our notion of terra firma, occurring in both time and space (sensu Ingold 1993) relative to one another and affected by both lunar gravity (the tides) and seasonal changes in ocean currents as well as the ability of people to engage with those interfaces and their sequelae. One of the most significant creations of human mind and hand for enabling such engagement in these contexts were various forms of watercraft, reported to be composite in design by early historic sources such as Ulloa, Vizcaíno, Taraval,

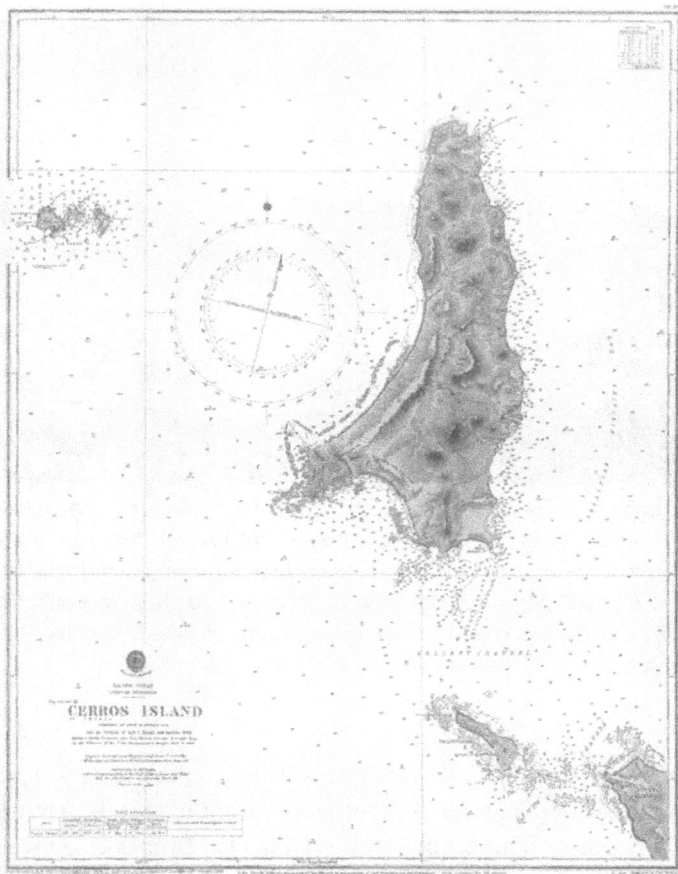

Figure 5.2. An exceptional example of late colonial cartography: British Admiralty Chart of Isla Cedros, Baja California: Isla Cedros, Mexico. Surveyed by Capt. H. Kellett, R.N., and the Officers of HMS *Herald* and *Pandora,* 1846. Kellett & Dewey Channels, the San Benito Islands & South Bay by the Officers of the USS *Narragansett* and *Ranger,* 1873 and 1889. Scanned by the authors from a paper copy of the chart.

and del Barco, given the dearth of large trees on much of the peninsula only marginally ameliorated by driftwood carried many miles and deposited along the coastlines by the California current (Des Lauriers 2005).

Relevant too is the observation that social networks identified on the peninsula documented through obsidian exchange patterns (Panich et al. 2015) consistently reveal an east–west axis of exchange, rather than north–south movement. These patterns are further reinforced by the distribution of various languages (e.g., Mixco 2010) also largely in a layered pattern all along the

Figure 5.3. English pirate Woodes Rogers landing on the coast of southern Baja California (from a print in the Macpherson Collection), likely in the Cape Region based upon the form of the native raft shown in the center of the engraving. This image was reproduced in a 1928 printing by Longmans, Green and Co., New York, of Woodes Rogers' book *A Cruising Voyage Around the World,* originally published in 1712. https://archive.org/details/cruisingvoyagero00roge_0/page/208/mode/2up.

peninsula connecting the two coastlines. This would strongly suggest that the social connections formed by communities of the central peninsula were typically between groups occupying overlapping landscapes of knowledge that encompassed the full range of variation experienced on the peninsula, in terms of alternate coastlines, elevation, and microclimates (e.g., Wilken-Robertson 2018). Further north, seasonal movements resulted in single mobile "communities" occupying alternating coasts depending on patterns of resource availability (e.g., Moore 1999, 2001). In either instance, we conclude that the variations were matters of choice rather than of restricted knowledge of various potential types of craft. The distribution of different designs (Figure 5.3) may have also shifted through time, with increasing aridity decreasing the availability of reeds for tule balsas in all but the most unusually wet locales. Also, with many of the larger oases of the peninsular desert being located at some distance from the coastline (e.g., Falcon-Brindis and León-Cortéz 2023), the mismatch between locations with abundant reeds and the bodies of water where craft manufactured from such would be employed presented another complication.

This brings up an interesting point about how our perception of the past struggles with the palimpsest of information scattered like shreds of a manuscript across the field of view afforded by incompletely distinguished time and space. Some landscapes were accessible without watercraft 15,000 years ago (e.g., Isla Cedros; Des Lauriers 2010) but today are located 23 km from the nearest point of the mainland or more, given continued sea-level rise. For the Baja California Peninsula, paleoclimatic research suggests that 15,000 years ago precipitation rates would have been higher than today, with some interior lake basins forming (Davis 2003) and marshy freshwater zones existing even on some of the landscapes that today form offshore islands. As such, an interesting progression would have faced the earliest inhabitants: As the land became increasingly arid, and terrestrial resources became either more sparse or harder to process, the seas rose simultaneously, creating increased incentive for the continued use of watercraft and emphasis on marine resources among many groups.

Even more counterintuitively, as precipitation declined further, the very tides would have become even more indispensable as a source of life-giving water in a fascinating interplay of hydrology and lunar gravity (see Cartwright 1999). The high evaporation rates prevailing through much of the Baja California Peninsula mean that in recent millennia, water sitting on the surface rapidly dries out (e.g., de las Heras et al. 2014). Small springs and even streams often have white crusts rimming their margins, even when the water itself is not particularly saturated with mineral salts. At the shore, when water flows down sandy-bottomed arroyos, it is protected from evaporation by flowing underground. As it reaches the shoreline, the freshwater—being less dense than the seawater—"floats" on top of the seawater, which intrudes the sediments near the shoreline, leading to greater accessibility for people at the water's edge. Wells were thus dug in the sandy beaches at the mouths of larger arroyos or canyons just above high tide line and were called *batequis* (see Aschmann 1959: 58–59). Additionally, the marine layer of fog (Figure 5.4) along the western coast provides a significant boost to the water budget of plants (see Rundel 1978) living within its range along the coastal slope and is literally life-giving in some of the more arid stretches of the peninsula (e.g., Vanderplank and Ezcurra 2016). The connections between these factors and shamanic power will be mentioned later.

It is likely that the earliest arrivals to the peninsula were "people of the sea," possessed with the full panoply of maritime technologies (Figure 5.5), including at least marginally seaworthy boats (see Des Lauriers et al. 2017). Little opportunity existed to develop such technology locally—especially given the relatively impoverished raw materials available for their construction in the central peninsula itself—and the specialized knowledge about necessary design features was almost certainly born into the space by migrating peoples who already knew how to build such craft.

Figure 5.4. Two peaks of Isla Cedros as viewed from the air. Both from the air and on the ground, the fog is thick, curtailing visibility and drenching every exposed surface with heavy mist. Photo by Matthew Des Lauriers.

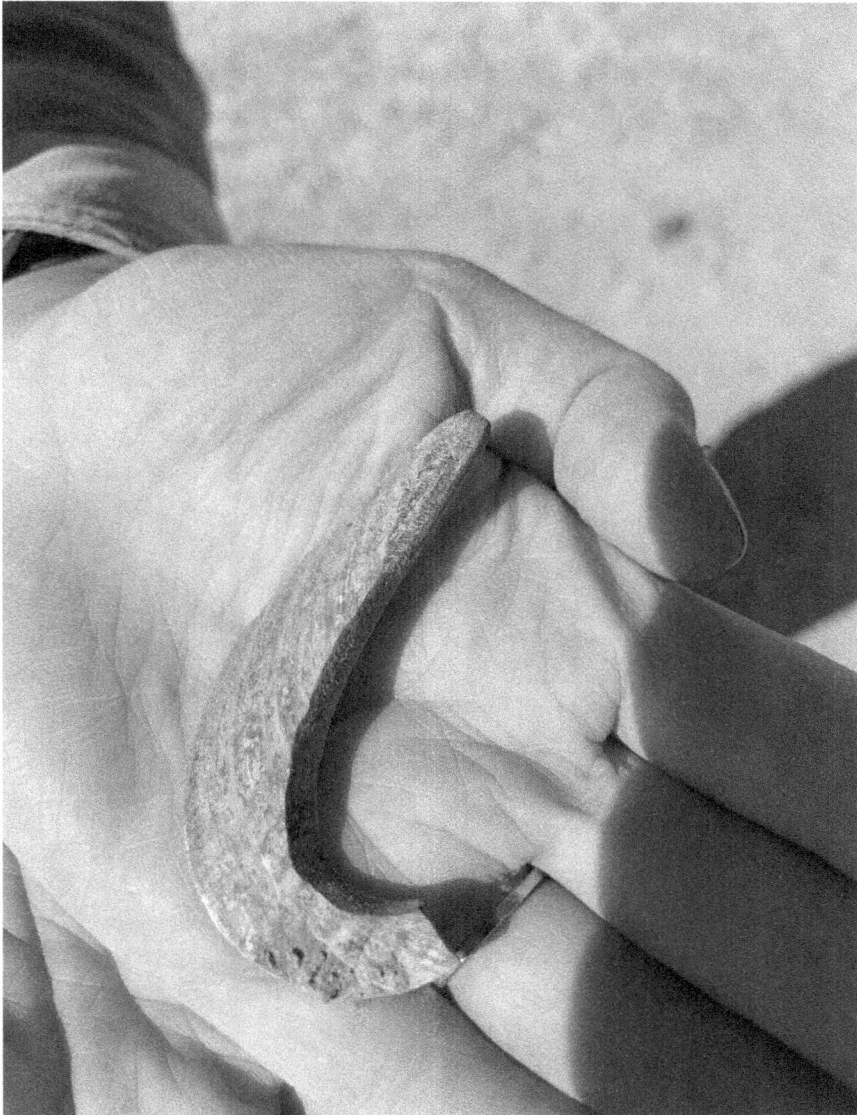

Figure 5.5. One of the larger Terminal Pleistocene / early Holocene shell fishhooks from Isla Cedros. Along with a range of hook sizes and types, nets and harpoons were also used by the inhabitants of the island from the very earliest periods of occupation. Photo by Matthew Des Lauriers.

Figure 5.6. Two examples of resilient and widely appropriated boat designs: the Indigenous North American Arctic kayak (*left*) and the ancient Mediterranean galley (*right*). *Left*: Inuk with a kayak being watched by another man and three women [Inuit] at Cape Dorset [Kinngait], Nunavut, 1948. Photo Credits: *Left*: S. J. Bailey, Department of Indian and Northern Affairs, Library and Archives Canada, e002213327. *Right*: Photo by Hugh Llewelyn, Flickr, Licensed under CC BY-SA 2-0, https://creativecommons.org/licenses/by-sa/2-0/.

This is significant not so much for issues of time depth but because it demonstrates the persistence and resilience of major physical materializations of knowledge and identity, even when migration and climate change create widescale landscape transformation. Boat designs are one of those elements of material culture that seem to remain remarkably consistent when their cognitive archetypes are borne along with migrating populations—for example, the widely dispersed Polynesian designs (Van Tilburg et al. 2018), Irish Currachs (Mac Cullagh 1992), or the North American Arctic umiaks and kayaks (Whitridge, this volume: Chapter 10). In modern times the near-global dispersal of the basic concept of the Indigenous kayak is an example of a resilient design adapted and appropriated widely. It would seem that the phenomenon of expansive transcultural sharing of technological design was less frequent in pre-globalization contexts, although the ancient Mediterranean galley (e.g., Olaberria 2014; Özbey 2019) could be argued as another example of such cross-cultural adoption of basic design principles (Figure 5.6).

There is at least one common factor for all of the various composite craft used along the expansive Baja California Peninsular coastline: They were designed to transport people rather than cargo. They allowed people to access offshore resources and move from one area to the next but were not really designed to carry items in bulk quantities in the same way that the watercraft of the Alta California Channel Islands (Arnold 2007; Hudson et al. 1978) or the Northwest coast of North America (Durham 1955; Neel 1995; Rorabaugh, this volume: Chapter 3;

Figure 5.7. Human and avian fishermen heading out to fish in the light of dawn. Isla Cedros, Baja California. Photo by Matthew Des Lauriers.

Smith, this volume: Chapter 9). For the people of the Baja California Peninsula, stability, durability, and resilience (given a short supply of raw materials to make replacements) were priorities over speed, cargo capacity, and the production of large numbers of vessels. Some models would have been relatively slow and not very graceful but basically unsinkable (see Des Lauriers 2005). They would have had low cargo capacity and a low profile, reducing the influence of wind on their speed (which was not much to begin with) and navigational direction. On Isla Cedros, these craft were described as displaying modularity in their design, with multiple craft combined to transport more than 12–15 people at a time, then separated to work as smaller fishing craft or for transporting 6–8 people (Des Lauriers and García-Des Lauriers 2006: 138). Consistently in the ethnohistoric literature for the peninsula (e.g., Baegert 1942 [1772]; del Barco 1988 [1776]; Venegas, in Mathes 1979), watercraft were essential tools for labor performed in the sea or moving people to places separated by water. In every known instance, it appears that design priority was given to supporting human activities in or on the water rather than transporting surplus goods.

A boat is a remarkable form of transportation that surpasses even domesticated animals for its cognitively transformative nature (Davidson 2010). A horse is wonderful, a horse is fast, a horse can carry far more than a person can. But at the end of the day, a horse still walks on the same ground that we walk. A boat takes us to places further than any human could ever swim. Until the development of powered flight, nothing else built by humans surpassed boats from that fundamental perspective. Even locomotives required arduously built tracks, which forever dictated their path with absolute rigidity. Compare that to the freedom of navigation of a boat on the water (Figure 5.7).

Boats are about going somewhere and doing something once you are there. The canonical anthropological concept of the Kula ring springs to mind (Leach and Leach 1983). Little emphasis is placed on the boats in terms of how the concept is often taught and discussed, but this is a huge oversight since it is the boats and not the *mwali* and *soulava* that truly circulate along and between those tropical shores. Rather than remaining as passive observers and receivers of the ocean's tempest, the watercraft of the Indigenous Baja Californians enabled them to engage with the watery domain on their terms and in ways that could not have happened otherwise. A partial list of examples includes the observation that many of the fishing communities of the Peninsula—both past and present—take advantage of steep bathymetry and the layer cake of seawater temperatures (thermocline) to avoid having to relocate their fishing activities in a two-dimensional sense. Rather than following sea surface temperatures and relocating their communities, they simply fish(ed) in the deeper water where the local cool-water-favoring fish retreat during the summer months when tropical currents from the south bring significantly warmer water northward. Additionally, boats would have been the only practical way to deploy nets to catch both very small and very large fish that could not be caught on hook and line. Evidence for the successful harvesting of both medium (~2–3 m long) sharks and large groupers (~250 kg) is clear evidence for the use of nets to harvest large prey (Des Lauriers 2010). Meanwhile, the abundance of remains from small fish like smelt in other deposits, which demonstrates the capacity to conduct the mass harvest (Madsen and Schmitt 1998) of fish that would not have been worth pursuing on individual levels. Both of these activities are facilitated by using stable, functioning watercraft, without which the feasibility of either activity drops significantly.

Another pattern that holds true today as well as in the past is the phenomenon of redistributing people between communities as the boom-and-bust cycles of landscape and resources created greater abundance or scarcity. Today, fishing camps in central Baja California are occupied during lobster and abalone seasons (September to February and December to June, respectively), but during the offseason, these locales are virtually abandoned, except for a small crew of caretakers and security personnel. All along the Baja California coastline, remote offshore islets, incapable of supporting permanent occupation due to small size or lack of water, were visited by Native people consistently for thousands of years. These places may have been ideal to exploit relatively pristine shellfish beds or sea mammal rookeries or may have had terrestrial resources that were less available or accessible elsewhere. They may also have been locations with profound sacred qualities, and the experience of standing on some of these remote, stark, windswept places does nothing to dissuade us from proposing this possibility.

Of massive social importance was the maintenance of social ties between communities and the sharing of news and information (Whallon 2006). This was evident given that news reached Isla Cedros in 1728 (Des Lauriers and García-Des Lauriers 2006), within months (if not weeks) of the founding of Mission San Ignacio on the mainland. Also illustrative is that, in 1540, the first contact between the crew of Francisco de Ulloa (Figure 5.8) and the Cedros islanders was a violent one (Montané Martí 1995), which strongly suggests that the locals had advance information about their Spanish "visitors." The nearest that Europeans had come to Cedros up to 1539 was the Bay of La Paz, almost 700 km distant "as the crow flies." Add to this that the Europeans had arrived at the Bay of La Paz less than four years prior. Clearly, the networks of information exchange that operated between the Indigenous communities of Baja California at the time of the conquest were robust and moved news fairly rapidly. But subsistence pursuits by themselves are not the most socially significant dimension of watercraft. The strength and frequency of intercommunity ties on the peninsula is borne out by numerous examples of translocated "exotic" goods and materials but is more clearly displayed by the rapid spread of information, by kinship ties linking widely separated groups, by consistent reports of intercommunity violence, and by whole families "visiting" and spending months at a time with relatives in other communities (see Venegas, in Mathes 1979). These are some of the motivations that were moving people across the land- and seascapes of pre-sixteenth-century Baja California.

Among many of the peninsular societies, there was a clear connection between smoke, clouds, fog, and generative or shamanic power. The Milky Way is seen as the smoke from the pipe of the sleeping Creator deity, who sang and dreamed the world and all its pathways into existence (Zazueta 1978: 98). Supernatural beings transform themselves into clouds and travel about in that form (Mixco 1983), and Isla Cedros bore the name Huamalgua, which is likely an intentionally polyvalent word drawing on the connection between shamans and fogs. In several cosmological accounts from Native peoples of

Figure 5.8. A Replica of the *San Salvador,* Joao Cabrillo's ship from a 1542 expedition that followed Ulloa's two years after first contact on Isla Cedros. This ship gives a very good idea of the type and size of vessel that would have carried the first Europeans seen on Isla Cedros by the native Huamalgüeños (Montané-Martí 1995). Photo taken in Morro Bay, California, in August 2023 by Elwood P. Dowd. Source: Wikimedia. Licensed under CC BY-SA 4-0, https://creativecommons.org/licenses/by-sa/4-0/.

Baja California, supernatural beings and "gifted" people could not only deploy various forms of fog, haze, and mist as the vehicle for enchantments but could actually transform themselves into clouds and other atmospheric phenomena (e.g., Mixco 1978; Zazueta 1978: 57). This is seen in many oral traditions around the world, such as the Lakota concept of *atmokinesis* to create obscuring and confusing fogs called a "Wolf's Day" to aid during raids and battles (Dorsey 1894), or the Irish *feth fiada* (magic fog) of the Tuatha Dé Danann (Joyce 1968), or the classic trickster hero Old Man Fog among Australian Aborigines of the Cape York Peninsula (Haviland and Hart 1998).

Looking at the Baja California cosmological geography, it is almost certain that many of the offshore islands of the peninsula were conceived of as sacred mountains (see Mixco 1983), lying as they do toward the cardinal points of sunrise and sunset depending on the particular example. These mountains were believed to be the original "homes" of the quadripartite clan divisions observed among most of the peninsular tribal groups (Zazueta 1978). Even among related groups in southern Alta California, like the Kumeyaay of the San Diego region, the offshore islets known today as the Islas Coronados were seen as representing the "Western Mountain" (see various references in Shackley 2004).

For the central peninsula, this role was—without much doubt—assigned to Isla Cedros. The great mass of Isla Angel de la Guarda (see Figure 5.1) may have the eastern counterpart for its region, although the island-rich Sea of Cortez may have too many smaller islands for us to be as certain about "Eastern Mountains" as we are for those in the direction of the open Pacific. The association of the Raven or Shaman's clan with the "Western Mountain" among surviving groups of the northern peninsula (Zazueta 1978) should not be ignored, especially given that the Cochimí of the central peninsula used the term *Guama* or *Wama* to refer to the Jesuit priests, and in 1732 the Jesuit missionary Miguel Venegas (Mathes 1979) recorded the name Huamalgua (sometimes written Guamalgua) for the island, with the suffix -*Wa* consistently being the locative—literally "-house" (Mixco 1978). He translated that word as meaning "Isla de Nieblas" or Island of Fogs (see Des Lauriers 2010; Des Lauriers and García-Des Lauriers 2006). However, a close examination of Jesuit records of the Cochimí language (which no longer has any living speakers) does not support this translation (see Mixco 1978). In fact, it is likely to have been more directly translated as "house of the shamans," if the nearly identical words used by central Baja California peoples to refer to the Jesuit priests and their own religious leaders are any indication. The distinct possibility exists that a polyvalent meaning for the original toponym was "lost in translation," or even intentionally withheld from the missionaries.

The physical experience of "land's end" from a human viewpoint is much more apparent along the Pacific margin of the peninsula than in the narrower, bounded sea between mainland Mexico and Baja California. When one thinks of a cloud-shrouded mountain sitting on what empirically appears to be the edge of the world (Figure 5.9), it is less difficult to imagine the process of mythologizing its landscape. It is not often that one has to get into a boat to reach the peak of a sacred mountain, but embedded within the cognitive landscape of sea-bounded Baja California Peninsula, this has a certain conceptual coherence.

Figure 5.9. Isla Cedros, Baja California, as seen from a small boat approximately 10 km offshore near the middle of the channel between the island and the peninsular mainland. Photo by Matthew Des Lauriers.

The unbroken ocean horizon also represented the "edge" between the knowable and unknowable. In several of the oral traditions of societies native to Baja California—including the Kiliwa—the supernatural being that was the embodiment of the Sun (I intentionally avoid the use of the term *God* in this context) is killed by a stingray as he enters the western ocean but is brought back from the dead, out of the waves (from the east) when he washes ashore and his face is licked by his dog (Mixco 1978; Zazueta 1978). The horizon line of the Californian Seas appears to have been the boundary of the "real" world, beyond which time and space were not subject to the same rules. This is seen in many societies, including ancient Irish mythologies of Tír na nÓg (Smyth 1998), the Japanese tale of Urashima Tarō (Holmes 2014), and even Homer's *Odyssey* (1996). It defines the edges of true empiricism since one is always at the center of one's own horizon, whether at sea or in deep space (DeGrasse Tyson and Trefil 2021; Halpern 2012). These were people who lived with a nearly omnipresent reminder of the unknown—but an unknown upon which they depended, and only their watercraft enabled them to gain even the smallest of entries into its vastness.

It would be a grave mistake to assume that the watercraft painstakingly constructed by these people were mere objects of no special importance. The amount of labor that would have gone into their construction, their use through the collective physical action of launching and paddling, and their capacity to transform human capabilities would have imbued them with meaning far beyond that of nearly any other thing created by human mind and hand. Given the deep links between boats and people and the seas that Jacques Cousteau called the "world's aquarium," it seems likely that, as John Steinbeck wrote in the *Log from the Sea of Cortez* (1951) "a boat, above all other inanimate things, is personified in man's mind." Boats, then, might be seen as actual members of these communities of maritime hunter-gatherers. An example of how transformative the reciprocal relationships between the artists and their "creations" can be when those objects or technologies are so central to the lives of their communities.

References Cited

Arnold, Jeanne E. 2007. Credit Where Credit Is Due: The History of the Chumash Ocean-going Plank Canoe. *American Antiquity* 72: 196–209.

Aschmann, Homer. 1959. *The Central Desert of Baja California: Demography and Ecology.* Ibero-americana 42, University of California, Berkeley and Los Angeles.

Baegert, Jacobo. 1942 [1772]. *Noticias de la Peninsula Americana de California.* Translated by Pedro R. Hendrichs Pérez. First Spanish ed. Antigua librería Robredo de José Porrúa e hijos, México D.F.

Cartwright, David Edgar. 1999. *Tides: A Scientific History.* Cambridge University Press, New York.

Davidson, Iain. 2010. The Colonization of Australia and Its Adjacent Islands and the Evolution of Modern Cognition. *Current Anthropology* 51(S1): S177–S189.

Davis, Loren G. 2003. Geoarchaeology and Geochronology of Pluvial Lake Chapala, Baja California, Mexico. *Geoarchaeology* 18(2): 205–223.

de las Heras, Alejandro, Mario A. Rodriguez, and Marina Islas-Espinoza. 2014. Water Appropriation and Ecosystem Stewardship in the Baja Desert. *Change and Adaptation in Socio-Ecological Systems* 1(1): 63–73.

DeGrasse Tyson, Neil, and James Trefil. 2021. *Cosmic Queries: StarTalk's Guide to Who We Are, How We Got Here, and Where We're Going.* Simon and Schuster, Chicago.

Del Barco, Miguel 1988 [1776]. *Historia Natural y Crónica de la Antigua California [adiciones y correcciones a la noticia de Miguel Venegas],* edited by Miguel León Portilla. Universidad Nacional Autónoma de México, Instituto de Investigaciones Históricas, Ciudad de México.

Des Lauriers, Matthew R. 2005. The Watercraft of Isla Cedros, Baja California; Variability and Capabilities of Indigenous Seafaring Technology along the Pacific Coast of North America. *American Antiquity* 70: 342–360.

Des Lauriers, Matthew R. 2010. *Island of Fogs: Archaeological and Ethnohistorical Investigations of Isla Cedros, Baja California.* University of Utah Press, Salt Lake City.

Des Lauriers, Matthew R., Loren G. Davis, John Turnbull, John Southon, and R. E. Taylor. 2017. The Earliest Shell Fishhooks from the Americas Reveal Fishing Technology of Pleistocene Maritime Foragers. *American Antiquity* 82(3): 498–516.

Des Lauriers, Matthew R., and Claudia García-Des Lauriers. 2006. The Huamalgüeños of Isla Cedros, Baja California, as Described in Father Miguel Venegas' 1739 Manuscript *Obras Californianas*. *Journal of California and Great Basin Anthropology* 26: 123–152.

Dorsey, James Owen. 1894. *A Study of Siouan Cults*. Smithsonian Institution, Bureau of Ethnology Annual Reports 11, pp. 351–553. Washington, D.C.

Durham, George. 1955. Canoes from Cedar Logs: A Study of Early Types and Designs. *Pacific Northwest Quarterly* 46(2): 33–39.

Edney, Matthew H. 2019. *Cartography: The Ideal and Its History*. University of Chicago, Chicago, Illinois.

Falcon-Brindis, Armando, and Jorge L. León-Cortés. 2023. The Oases of Baja California Peninsula: Overlooked Hotspots for Wild Bees. *Journal of Insect Conservation* 27: 117–128.

Halpern, Paul. 2012. *Edge of the Universe: A Voyage to the Cosmic Horizon and Beyond*. Turner Publishing, Nashville, Tennessee.

Haviland, John B., and Roger Hart. 1998. *Old Man Fog and the Last Aborigines of Barrow Point*. Smithsonian Institution Press, Washington, D.C.

Holmes, Yoshiko. 2014. Chronological Evolution of the Urashima Tarō Story and Its Interpretation. Master's thesis, School of Languages and Cultures, Victoria University of Wellington, Wellington, New Zealand.

Homer. 1996. *The Odyssey*. Translated by Robert Fagles. Penguin, New York.

Hudson, Travis, Janice Timbrook, and Melissa Rempe (editors). 1978. *Tomol: Chumash Watercraft as Described in the Ethnographic Notes of John P. Harrington*. Ballena Press, Socorro, New Mexico.

Ingold, Tim. 1993. The Temporality of the Landscape. *World Archaeology* 25: 152–174.

Joyce, Patrick W. 1968. *A Social History of Ancient Ireland*. 2 vols. Benjamin Blom, New York.

Leach, Jerry W., and Edmund Ronald Leach (editors). 1983. *The Kula: New Perspectives on Massim Exchange*. Cambridge University Press, Cambridge.

Mac Cullagh, Richard. 1992. *The Irish Currach Folk*. Wolfhound Press, Dublin.

Madsen, David B., and Dave N. Schmitt. 1998. Mass Collecting and the Diet Breadth Model: A Great Basin Example. *Journal of Archaeological Science* 25: 445–455.

Mathes, Michael W. (editor). 1979. *Obras californianas del padre Miguel Venegas, SJ*. Universidad Autónoma de Baja California Sur, La Paz, México.

Mathes, Michael W. 1989. The Mythological Geography of California: Origins, Development, Confirmation and Disappearance. *The Americas* 45: 315–341.

Maxwell-Stuart, Peter G. 1981. *Studies in Greek Colour Terminology: Charopos*. Brill, Leiden.

Mixco, Mauricio J. 1978. *Cochimi and Proto-Yuman: Lexical and Syntactic Evidence for a New Language Family in Lower California*. University of Utah Press, Salt Lake City.

Mixco, Mauricio J. 1983. *Kiliwa Texts: "When I Have Donned My Crest of Stars."* University of Utah Press, Salt Lake City.

Mixco, Mauricio J. 2010. Las Lenguas Indígenas. In *La Prehistoria de Baja California: Avances en la Arqueología de la Península Olvidada*, edited by Don Laylander, Jerry D. Moore, and Julia Bendimez-Patterson, pp. 31–52. Centro INAH Baja California, Mexicali, Baja California, México.

Montané Martí, Julio C. 1995. *Francisco de Ulloa: Explorador de ilusiones*. Alforja del Tiempo, Universidad de Sonora, Hermosillo.

Moore, Jerry D. 1999. Archaeology in the Forgotten Peninsula: Prehistoric Settlement and Subsistence Strategies in Northern Baja California. *Journal of California and Great Basin Anthropology* 21: 17–44.

Moore, Jerry D. 2001. Extensive Prehistoric Settlement Systems in Northern Baja California: Archaeological Data and Theoretical Implications from the San Quintín-El Rosario Region. *Pacific Coast Archaeological Society Quarterly* 37: 29–51.

Neel, David. 1995. *The Great Canoes: Reviving a Northwest Coast Tradition*. Douglas & McIntyre, Vancouver.

Olaberria, Juan-Pablo. 2014. The Conception of Hull Shape by Ship-Builders in the Ancient Mediterranean. *International Journal of Nautical Archaeology* 43: 351–68.

Özbey, Muhammet Talha. 2019. Geç Dönem Ortaçağ'dan Altın Yelken Çağının Başlangıcına, Akdeniz'de Gemi İnşa Ve Dizaynına Dair Yazılı Tezler (İS. 1300–1650). *Çanakkale Araştırmaları Türk Yıllığı* 17: 255–271.

Panich, Lee M., Érika Moranchel Mondragón, and Antonio Porcayo Michelini. 2015. Exploring Patterns of Obsidian Conveyance in Baja California, Mexico. *Journal of California and Great Basin Anthropology* 35: 257–274.

Rundel, Philip W. 1978. Ecological Relationships of Desert Fog Zone Lichens. *Bryologist* 81(2): 277–293.

Shackley, M. Steven (editor). 2004. *The Early Ethnography of the Kumeyaay*. Classics in California Anthropology, Phoebe A. Hearst Museum of Anthropology, University of California, Berkeley.

Smyth, Daragh. 1998. *A Guide to Irish Mythology*. 2nd ed. Irish Academic Press, Blackrock, Co. Dublin.

Steinbeck, John. 1951. *The Log from the Sea of Cortez*. Viking, New York.

Van Tilburg, Hans K., David J. Herdrich, Michaela E. Howells, Va'amua Henry Sesepasara, Telei'ai Christian Ausage, and Michael D. Coszalter. 2018. Row as One! A History of the Development and Use of the Samoan Fautasi. *Journal of the Polynesian Society* 127(1): 111–136.

Vanderplank, Sula E., and Exequiel Ezcurra. 2016. Marine influence controls plant phenological dynamics in Mediterranean Mexico. *Journal of Plant Ecology* 9: 410–420.

Whallon, Robert. 2006. Social Networks and Information: Non-"Utilitarian" Mobility among Hunter-Gatherers. *Journal of Anthropological Archaeology* 25(2): 259–270.

Wilken-Robertson, Michael. 2018. *Kumeyaay Ethnobotany: Shared Heritage of the Californias*. Sunbelt Publications, San Diego, California.

Zazueta, Jesús Ángel Ochoa. 1978. *Los Kiliwa y el Mundo se Hizo Así*. Instituto Nacional Indigenista, México.

6

Kanči

Indigenous Seafaring, Watercraft Diversity, and Cultural Contact in Southern Patagonia

NELSON AGUILERA ÁGUILA,
ALBERTO GARCÍA-PIQUER,
AND RAQUEL PIQUÉ

When the first Europeans arrived in Patagonia and Tierra del Fuego in the sixteenth century, the region was occupied by a diversity of Native peoples who were classified in a very oversimplified manner based on major physical attributes and their way of life. There were land groups in or close to the steppe regions of continental Patagonia, called "Patagones." These were the famous giants of the first Spanish reports. And then there were other people, the canoe peoples, who remained invisible to the European eye for several decades. Later they would be called "Pecharai," but little was known about them. In fact, in a large part of the Patagonian waters, contact between Native peoples and Western sailors was rather occasional until well into the eighteenth century. It was not until the next century that growing scientific and ethnographical interest led to the identification and partial study of six ethnic groups distributed from the 41° parallel to the south (Fitz Roy 1839). For three of these peoples, the canoe was central to the way of life, and they were called by the ethnonyms Chono, Alacaluf (currently known by the autonym Kawésqar, which is the term adopted in this chapter), and Yagán (Figure 6.1).

Due to the singularity of the colonization process, European contact did not substantially impact the way of life of Patagonian peoples until the nineteenth century. However, it generated a very rich ethnohistorical record, both in quantity and quality. Therefore, the archipelagos and channels of Patagonia provide a unique opportunity for the study of Indigenous seafaring and native boat-building technologies.

Patagonia extends from parallel 41°30′S to Cape Horn (parallel 56°S), with a surface area that exceeds 1.5 million sq. km. As can be expected, this huge

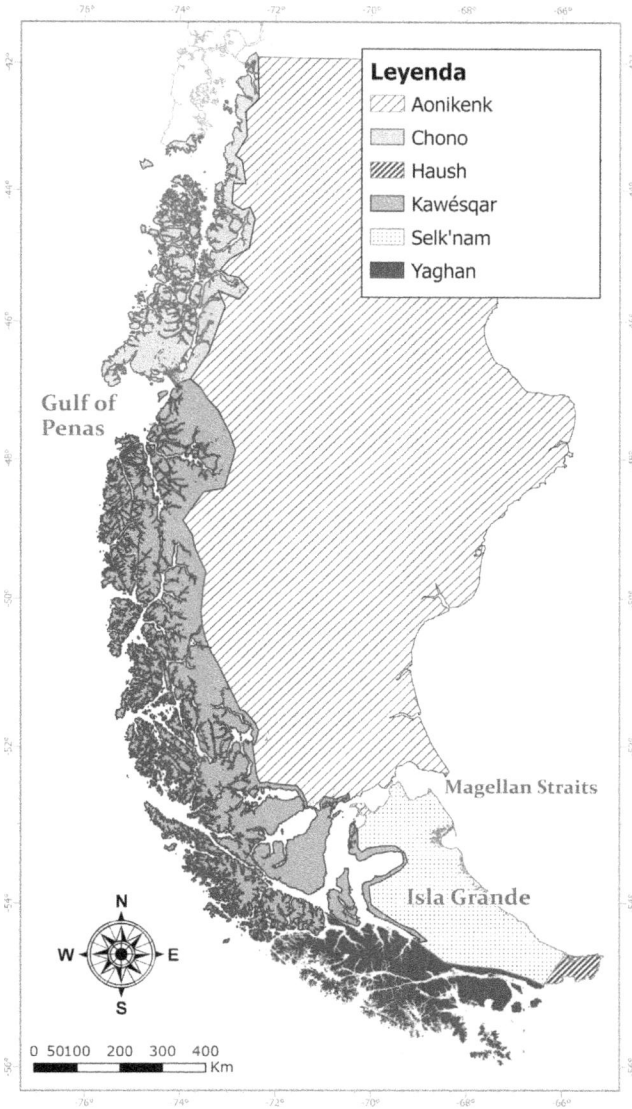

Figure 6.1. Ethnic distribution in historical times in Patagonia.

macroregion is highly diverse in geographical and ecological terms, presenting sharp internal contrasts. The study area can be divided longitudinally, using the Andes Mountains as a landmark, or latitudinally, considering the Taitao Peninsula and the Strait of Magellan as boundaries since they represent transitions between different ecological zones. Each region offers particularities resulting from large-scale environmental dynamics, which represent nominally homogeneous scenarios for human occupation since its first evidence 12 millennia ago. Studying the history of watercraft in the region could be also helpful to better understand how the different seafaring peoples interplayed with specific environments, and to assess the social and cultural variation within those watery worlds.

Charles Darwin (1840) noted in 1830 that canoes (referring to bark canoes) were probably the most ingenious invention of the Fuegian people, which indicated that the canoes had not changed in the previous 300 years. Indeed, the bark canoe was considered as a stable technological solution, the consummation of native craftiness achieved probably thousands of years earlier. However, there are many observations by Western navigators providing evidence that canoe-building was rather a highly dynamic field. Ethnohistorical sources recorded three main types of Indigenous watercraft in Patagonia (Figure 6.2). The first of these were the bark canoes seen in the Strait of Magellan and the Fuegian channels. The second corresponded to the plank canoes, called *dalkas*, observed mainly in the northern area of Patagonia. The third category is dugout canoes. Lothrop (1932), citing Lord Byron's observation of *dalkas* in the Strait of Magellan [1765], wondered if there were no earlier records of this type of canoe in the area. The question is important because plank canoe technology originated 1,200 km to the north. Several researchers have analyzed this type of craft and have proposed hypotheses that point to cultural diffusion or contact between neighboring ethnic groups. Homogeneity was the exception rather than the rule, with each type of watercraft presenting variations in size, construction techniques, use, and materials employed for its manufacture, which would imply that technological stability hides a transition whose origin we are now exploring (Aguilera 2023).

The observation of Indigenous navigation is not absent in ethnological accounts. However, classic nineteenth- and early-twentieth-century works overlook this aspect of the life of the canoe peoples. They are defined as nomadic seafarers but described as if they were pedestrian peoples. There are rather exhaustive descriptions of the types of dwellings they built and the activities they were involved in on the shore, but we know little about how life on board boats was, or how their travels developed and the reasons behind the use of a specific type of canoe, given the diversity of types. Canoes are a diagnostic and meaningful element of nomadic maritime cultures—diagnostic because they define a way of life and a way of constructing cosmovision and identities that still survive in the perspective of the Indigenous people who knew this tradition and lived it personally; meaningful because it is an attribute of the material culture loaded with semantics, which defines a type of relationship with the world. This can be seen in particular forms of worldview, ethnogeographic understanding, and social interaction.

Figure 6.2. The three main types of Indigenous watercraft in Fuego-Patagonia. *Upper left*: *dalka* or plank canoe, Etnografiska Museet, Stockholm. Photo courtesy of Gabriel Zegers. *Upper right and down*: dugout and bark canoe held in the Museo Maggiorino Borgatello, Punta Arenas, Chile. Photos courtesy of Claudio Miranda Díaz.

In this chapter we focus on the emergence of new types of watercrafts in the region, with a particular interest in plank canoes. First, we review when and where each watercraft type was first recorded by ethnohistorical sources and how they spread to other regions. We use a spatiotemporal quantitative approach to the ethnohistorical sources to analyze the evolution of each type during the contact period (1520–1960). In the following sections, we discuss the hypothesis of cultural diffusion or contact proposed to explain the spreading of new types of watercraft, and we consider if the regional availability of forest and wood species played a role in defining the edges of the areas in which different types of canoes were used. Last, we consider the impact that these new types of watercraft could have had on the nautical practices and organizational strategies of Southern Patagonian and Fuegian seafaring peoples, traditionally built around the bark canoe.

Materials and Methods

The present work is based on the analysis of Patagonian ethnohistorical sources, including voyage chronicles, journals, exploration reports, and scientific works, with the aim of individualizing in time and space multiple observations of Indigenous peoples (often described in texts as "sightings"). This approach has been applied previously to demographic studies (Estay et al. 2022; Martinic 1989) and to the study of mobility and climatic adaptations in the region (Pallo 2011). However, these previous works used a dataset composed of 140 historical observations of Indigenous peoples (Martinic 1989). To expand this dataset, both chronologically and spatially, we have engaged in a new analysis of the ethnohistorical record of Patagonia, with a particular interest in southern Patagonia, reviewing a total of 1,299 sources. These sources correspond to 750 travels or explorations realized during the years 1520 to 1960 both in northern Patagonia (north to the Gulf of Penas; see Figure 6.1) and in southern Patagonia. Sources relating to southern Patagonia are much more abundant, representing 89% of the total.

For each sighting, the time of navigation, location, ethnicity of crew, and number and type of watercraft as well as qualitative data such as descriptions of seafaring equipment, raw materials, and craft techniques, has been recorded. It is important to note that there is not always a one-to-one correspondence between sightings and individuals. In some cases, certain individuals were sighted by multiple observers. This circumstance has been considered in order to minimize the duplication of cases.

Our database is therefore composed of quantitative and qualitative data. The data was collected, systematized, and compiled in a spatial database of the ESRI geodatabase type (*.gdb). Then it was analyzed spatially and temporally with quantitative and GIS-based methods. We used ArcGIS Pro 2.8 by ESRI in combination with the statistical language R to perform all the analysis. From its cartographic deployment, the information related to sightings of Indigenous boats and camps observed by European navigators was analyzed, which allowed us to observe the distribution of Indigenous occupation in the area. The complete description of the methods and techniques applied can be found elsewhere (Aguilera 2023).

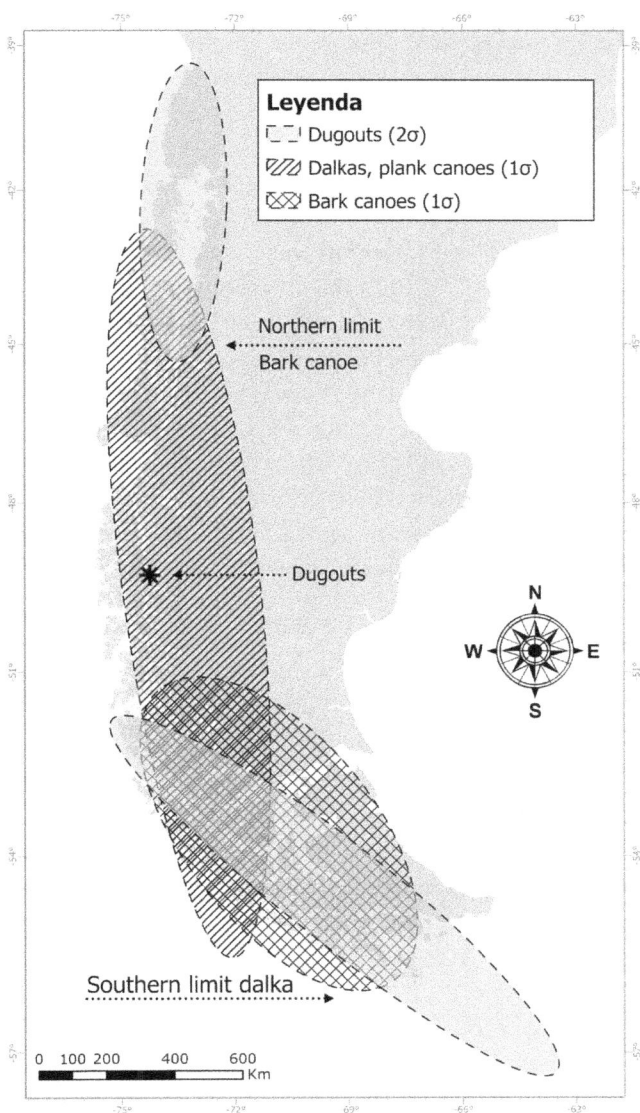

Figure 6.3. Distribution of the three main canoe types from the standard deviation ellipse: 1σ equals 68% of a normal distribution, 2σ equals 95%. For clarity, the two dugout canoe cores have been considered separately and an isolated observation (*) has been noted. The northern and southern limits, respectively, of the bark canoe and *dalka* are indicated.

Results

The dataset is composed of 2,260 observation events or sightings, from which 1,108 correspond to sightings including at least one watercraft. From these watercraft sightings, 221 (19.9%) were located in the Chiloé, Guaitecas, and Chonos archipelago area (northern Patagonia); 194 (17.5%) in the western channels area; 310 (27.9%) in the Strait of Magellan and inland seas; and 381 (34.4%) in the Fuegian archipelago.

From the total number of Indigenous watercrafts sightings that have been recorded in the sources ($N = 1.108$), the observers indicated the type of hull of the canoe in 397 cases (35.8%). Of these, 98 (24.6% of the known total) were bark canoes, 85 (21.4%) were plank canoes, 37 (9%) were dugout canoes, 4 (1%) were leather canoes, and 1 (0.3%) was a skin raft. We have excluded from analysis 170 additional observations of nonvernacular watercraft used by Indigenous or mixed crews (42.8%). These consist mainly of "piraguas," an eighteenth-century hispanized version of the *dalka*—larger, with rudder and sail—used exclusively in northern Patagonia. Overall, the distribution of types of watercraft is spatially and chronologically heterogeneous (Figure 6.3).

The Spatiotemporal Distribution of the Bark Canoe

The presence of the bark canoe was confirmed during almost all the study period, from the first canoe sighting in 1526 to the first decade of the twentieth century, particularly in southern Patagonia. This type of hull was dominant in the Strait of Magellan and especially in the Fuegian archipelago, where it tends to focus from the nineteenth century on. In the western channels, bark canoes coexisted with *dalkas*, once this type of hull began to be observed in the area, without being completely replaced until the beginning of the twentieth century. In the first decade of the twentieth century bark canoes virtually disappeared. The last sightings of these canoes in use correspond to Robert Mossman's account in 1905 (Rudmose et al. 1906). In northern Patagonia, there were only two sightings of bark canoes. The two known cases of the presence of bark hulls north of the 47° parallel suggest the maximum limit of expansion of this technology and correspond to episodes of mobility generated by the contact with Spanish missionaries coming from Chiloé (Urbina 2017).

The Spatiotemporal Distribution of the Plank Canoe

The most significant feature identifiable in our analysis is the spatiotemporal pattern presented by the plank canoe technology (Figure 6.4). Use of the *dalka* is, at first, clustered in northern Patagonia. Most descriptions place them in

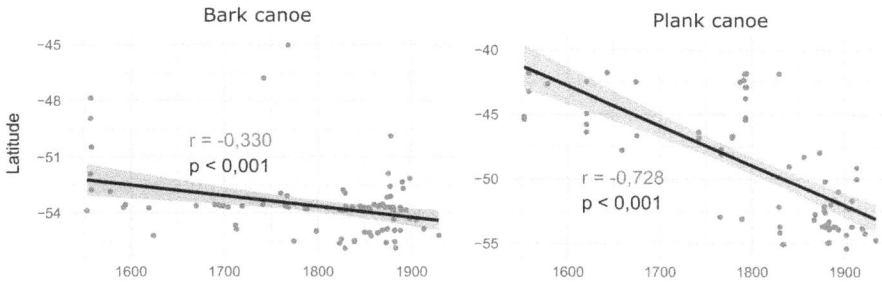

Figure 6.4. Correlation between time and latitude for bark (*left*) and plank (*right*) canoes. Correlation (r) and significance (*p*) values are indicated.

the Seno de Reloncaví (Medina 1923) and eventually in Guaitecas (Vidal 1880) during the sixteenth century. From 1600 onward, the missionary and military "entries" to the Guaitecas and Chonos regions made it possible to observe a greater number of this type of canoe. There are descriptions that locate them on the eastern coast of Chiloé (42°28'S/73°45'W), west of the Guaitecas Islands (44°56'S/73°37'W), and on Traiguén Island (45°40'S/73°48'W). During the eighteenth century, sightings of *dalkas* decreased in the northern area. In 1741 a group of survivors of the wreck of the *Wager* observed them in Guayaneco and subsequently used them to cross the Deshecho Pass (Urbina 2018).

There are no plank canoes south of the Gulf of Penas in this period. Crew members who fled into the Strait of Magellan in 1742 did not see any of these watercraft south of Guayaneco (Bukeley and Cummins 1743). However, this form of construction would eventually spread into southern latitudes in the next two decades. The first sighting of these watercraft is reported at Cape Upright, Strait of Magellan (52°56'S/74°10'W), in 1765 (Hawkesworth 1773). By the end of the eighteenth century, the tendency has been reversed: the *dalka* was scarcely seen north of the forty-eighth parallel, while it had become the dominant watercraft in the western channels.

The nineteenth century shows a high relative presence of *dalkas* in the Strait of Magellan. The sightings happened first in the Mayne Pass area (52°36'S/70°40'W; Bertrand 1886; Fitz Roy 1839; Lemay 1881; Mac-Érin 1897). During the twentieth century, sightings of *dalkas* were constant in the Western Patagonian Channels and in the Strait of Magellan. Between 1919 and 1922 some *dalkas* were observed south of Dawson Island (53°50'S/70°27'W; Kent 1924). The last sighting of this type of watercraft in use by Indigenous crews occurred at Burnt Island (54°43'S/71°12'W) in Desolada Bay in 1934 (Burg 1937).

The Spatiotemporal Distribution of the Dugout Canoe

Dugout canoes appear in the late stage of contact with Western sailors and set-tlers. The data point to the appearance of this type of watercraft in two places and at two times. The processes of adoption of the dugout canoe technology do not seem to be interrelated, taking place at least 100 years apart and occurring in a range of distance close to 1.450 linear km. We note the parallel distribution north of 46° and south of 49°S. Observations are concentrated in the Fuegian archipelago around 1870 (see Figure 6.3).

In northern Patagonia, the first observation of dugouts corresponds to a reference by Francisco Menéndez in 1780 (Urbina 2018) at the Cailín mission (43°11'S/73°33'W). José Moraleda (1796) sighted one of these abandoned canoes on Auchile Island (45°01'S/73°32'W), suggesting the incorporation of this form of naval architecture from northern techniques. This type of canoe continued to be sighted throughout the nineteenth century in the Chacao Channel and in the northern limit of the Reloncaví Sound (Hudson 1857; Maldonado 1915; Medina 1923).

At the end of the nineteenth century dugout canoes appeared in southern Patagonia. In October 1875 Pierre Reynaud (1876) reported a canoe of this type in Bahía Isthmus (52°9'S/73°35'W). In June 1876 Thomas Bridges (1930) reported one of these hulls at Courtenay Bay (54°37'S/71°20'W). Since 1880 these canoes began to replace traditional bark canoes in the Beagle Channel area, with occasional arrivals at the Anglican mission in Ushuaia, and finally they became dominant until the 1930s. In the western channels, dugout canoes were observed in Cutter Cove (53°22'S/72°25'W) in the first decade of the twentieth century (Martinic 1989). Since 1920 dugout canoes were observed frequently in this region, becoming dominant over a period of twenty years (Bird 1988; Legoupil and Chevallay 2017). They are the last type of Indigenous watercraft described in the area, with observations placing them in the Puerto Edén area in the late 1950s (Carabias 2019).

Discussion

The Spreading of the *Dalka:* Marker of a Migratory Process?

Direct archaeological finds related to Indigenous navigation in Patagonia are exceptionally rare and consist of fragments of a plank canoe or *dalka* found in the 1920s (Bird 1988) and pieces of canoes recorded by Nicolás Lira (2015, 2018; Lira et al. 2015) in northern Patagonia. Archaeological investigations carried out in southern Patagonia have rarely provided physical evidence of watercraft used in the past. The only exception is one fragment pertaining to a bark canoe recovered during an excavation on Navarino Island whose chronology is not clearly assessed yet (Bird 1988; Bird 2012; see Aguilera et al. 2021 for a reanalysis). The preservation conditions of the subsoil make it unlikely that organic remains like wood or bark materials would survive the passage of time. It has also not been possible to obtain information on the boatbuilding processes, and research on Indigenous shipyards is a topic that has not been developed in the region of Patagonia and Tierra del Fuego. It is possible that investigations on systematically excavated archaeological sites have not recognized in the evidence those attributes related to these areas of work. While underwater archaeology is still poorly developed in the region, future work on submerged context may prove fruitful.

Previous studies about watercraft and boat-building technology in Patagonia have drawn mainly from ethnographic monographs or from a small number of canoes found in various museums around the world that were collected in ethnographic contexts by explorers or missionaries (see Arnold 2017). Most of them are preserved in excellent condition and have been used to study technical aspects of their production, but they do not allow for more far-reaching conclusions because of their limited number and the vagueness of the finding information in most cases, with unknown location and year of finding (e.g., Carabias 2019). Moreover, especially regarding bark canoes, they are generally twentieth-century models, in a few cases probably crafted on demand for the ethnographers (Aguilera 2023).

Therefore—and despite the ethnohistorical record also having a relatively short scope, with the first watercraft sighting not happening until the year 1526—our approach opens a broader window into Indigenous canoes and their actual use. The results presented here confirm that the bark canoe was the unique type of watercraft used from south of the Gulf of Penas to Cape Horn when the first European contact occurred. Their reported distribution during sixteenth-century navigations locates them in the Strait of Magellan and the Fallos Channel. They were observed as far east as the eastern islands (52°51′S/70°38′W) and as far north as the Guayaneco Archipelago (47°53′S/75°17′W). It was the dominant canoe south of the Gulf of Penas since the first contact with European crews, and descriptions of its characteristics tend to be similar.

The *dalka* was generally attributed to the Chono tradition, a canoe culture of the south-central coast of Chile, which has been described as extinct since the early nineteenth century (Hudson 1857). The origin of this type of hull is unknown, but it was the canoe used in the Chiloé area when the Spaniards arrived at the end of the sixteenth century (Finsterbush 1934; Medina 1984; Puente 1986) and seems to be the result of the adaptation of wood technologies (sensu Rivas et al. 1999), a consequence of contact with continental Indigenous populations around 1000 CE (Aldunate 1996). These canoes have been described from the Bahía Concepción to the Gulf of Penas, but chronicles place them in the Strait of Magellan from 1765 (Lothrop 1932). Previous explanations for expansion have revolved around exchange with northern Chono communities (Emperaire 2014; Latcham 1930) or cultural diffusion (Finsterbush 1934; Lira 2015).

Based on the results obtained through the analysis of the ethnohistorical sources, we propose that the arrival of the *dalka* in the western channels is primarily the consequence of a small-scale migratory process of the Chono population from their traditional territory toward the south. Historical records (Lozano 1755; Techo 1897 [1673]) suggest that from 1710 onward, some Chono communities began to leave their traditional territories, escaping from Spanish slavery expeditions (Urbina 2017). Some of these groups are reported in the Gulf of Penas around 1741 (Byron 2017). Of course, the southward migration of Chono populations, and thus of their plank canoes, could have led to other cultural processes, such as intermarriage and cultural assimilation with the Indigenous communities of the western channels. In any case, plank canoes were used in the western channels by these Chono populations but were also adopted (or exchanged) by Kawésqar communities. However, the bark canoe was not completely replaced, as it continued to be used until the nineteenth century in the western channels.

The Dugout Canoe Innovation: Parallel Processes

The comparison of the spatiotemporal distribution of plank canoes and dugout canoes suggests two very different patterns. Dugout canoes would not be the consequences of one process but two independent ones. The first, in northern Patagonia, was presumably related to Hispanic and Mapuche influences on the natives of Chiloé. The second pattern, in southern Patagonia and Tierra del Fuego, was likely due to the Anglican missionary presence in the area and the adoption of metallic instruments.

There is no definitive evidence that indicates the exact geographical origin of the incorporation of this form of shipbuilding south of the forty-eighth parallel (Orquera and Piana 1999). Bridges identifies them as *luputuj anan,* large canoes of Alacaluf origin (Bridges 1930). Martin Gusinde (1986) also suggested that this type of hull was an incorporation from northern canoe communities (Kawésqar) who spread it to the Fuegian archipelago at the end of the nineteenth century. This conclusion was followed by some authors (Bridges 1952; Lothrop 1932), but the analysis of the sources does not allow conclusive judgments to be offered. The appearance of this type of watercraft coincides with the demographic collapse of the canoe communities in the channels of southern Patagonia and Fuegian archipelago, occurring in synchrony with the creation of permanent settlements in the area and the generalization of the use of metallic tools acquired by exchange.

There is a temporal and geographic hiatus between the two processes of incorporation of the dugout canoe in the study area. With the known dates, we know that a century elapsed between the introduction of these hulls in northern Patagonia and the Fuegian archipelago. An interesting observation is that dugout canoes were not observed in the Ofqui isthmus surroundings, an area that spanned two degrees of latitude. Although this could be an effect derived from a record bias, the historical analysis leads us to suppose that the weight of this type of hull would have been a handicap when trying to use the well-known portage route across the isthmus. Some attempts to portage dugouts ended up causing damage to the hulls (García 1889). Therefore, the structural factor could have played a limiting role in the expansion of this type of canoe southward, unlike what happened with the *dalka*, whose hull could be easily disassembled and transported in pieces.

The Forest and the Watercraft: Specialization and Flexibility in Wood Selection

The strong relationship between the use of specific types of watercraft seen in the ethnographic record might be correlated with the regional availability of different tree species, which may have led to different experiences and alternatives through time. In any event, the outcome of the process is clear. In the Northwest coast of North America, boats cannot be understood without the cedar (Ames 2002; see Smith, this volume: Chapter 9). In the opposite tip of the Americas, particularly in southern Patagonia, bark canoes are directly associated with the Fagaceae forests, especially *Nothofagus betuloides* as a key species (Figure 6.5). While the distribution of these forest communities is ubiquitous in the whole study area, the quality and abundance of individuals may vary. The highest and straightest specimens are found in protected valleys (Thiers and Gerding 2007). Indigenous parties traveled into the forest rarely, but when they did visit the forest, it was often in search of good trees for stripping bark to build canoes (Prieto et al. 2000). The dense *Nothofagus* forests are more easily found in the protected regions close to the continent, than on the exposed Pacific coasts. This fact could have played a key role in defining the settlement pattern and the mobility strategies of communities building bark canoes, particularly in the spring season when the bark was ready to be extracted.

A different case is that of the *dalkas*. In northern Patagonia, particularly on Chiloé Island, the use of *dalkas* has been frequently related to the exploitation of forest species such as the Guaitecas cypress (*Pilgerodendron uviferum*) or the larch (*Fitzroya cupressoides*). The height of their trunks and the inner structure of the wood would have allowed boat builders to extract long planks from these trees, similar to the ones needed for building the *dalka*. However, a close review of the ethnohistorical and ethnographic record (Aguilera 2023) shows that the range of raw materials used for crafting this type of canoe was much more diverse. According to the records, south to the Gulf of Penas, *dalkas* were composed mainly of planks obtained from Fagaceae such as *Nothofagus dombeyi*, *N. nitida*, and especially *N. betuloides* (Figure 6.5). These results coincide with the analysis of the few fragments of *dalka* that have been recovered (Lira 2015). Cypresses and larches could be found with relative frequency along the western channels, south of the Gulf of Penas up to the Strait of Magellan. However, in the southern latitudes climate, cypresses do not grow at the same annual rate that northern specimens do, which ultimately affects their height and diameter (Roig and Boninsegna 1991).

Contrary to what we observe in bark canoes and for dugout canoes (with a smaller sample), where there is a strong relationship between the type of watercraft and the presence of *N. betuloides*, the building of plank canoes shows high versatility in materials engineering. This flexibility could be one of the factors that explain the fast spread of this watercraft in the western channels (Aguilera 2023).

The Social Implications of Adopting New Watercraft

Regardless of the type of canoe we analyze, one aspect to keep in mind is that watercraft configure a small and closed world—small because its space is limited, and closed because the hull allows a maximum load capacity that is not expandable (volume and weight). It is a "machine" (sensu Bjerck 2016) that circumscribes the type and intensity of the distribution of space on board and the relationships that occur within it.

In bark canoes, the average lengths and capacities described suggest that the use of available space for people and objects was tacitly regulated. In a functional hull of more than 3 m in length, the internal spaces had to accommodate men, women, and children, each of whom had assigned functions and restrictions related to maintaining the stability of the watercraft (Bridges 1930; Gusinde 1986). Hein Bjerck (2016) has observed the existence of a strong "human–boat–settlement relation" in southern Patagonia, particularly in the Beagle Channel. This relation would have laid on the standardized size of the bark canoes, defining the size of the social unit and, consequently, the size of

Bark canoe

Plank canoe

Dugout

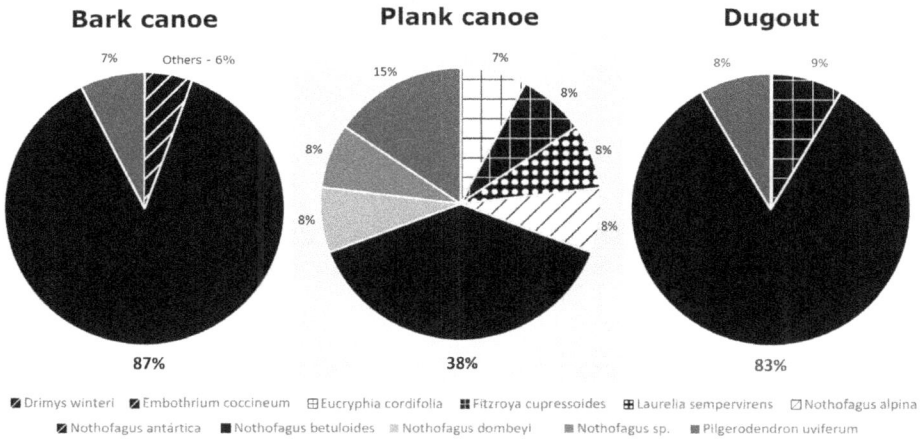

7% Others - 6%

8%

8%

87%

7%
15%
8%
8%

8%
8%
8%

38%

8% 9%

83%

🔳 Drimys winteri 🔳 Embothrium coccineum ⊞ Eucryphia cordifolia ▦ Fitzroya cupressoides ⊞ Laurelia sempervirens ▨ Nothofagus alpina
🔳 Nothofagus antártica ■ Nothofagus betuloides ▒ Nothofagus dombeyi ■ Nothofagus sp. ■ Pilgerodendron uviferum

Figure 6.5. Percentage of forest species used for each type of watercraft based on the ethnohistorical sources.

the dwelling area. As Bjerck himself remarks, this is one of multiple types of human–boat relations that can define a seafarer society.

Our analysis of the ethnohistorical sources supports this claim, indicating that the maximum number of adults in bark canoes remained in the range of four to five crew members and two to four children (Aguilera 2023; see Table 6.1). Rarely was it reported that more than two young men or women voyaged in the same canoe, which suggests that once they reached adolescence, young people were encouraged to build their own canoes, and young women of marriageable age were engaged in marriage. This factor could have contributed to population size control and bride exchange rules in a decisive manner since the bark canoe offered carrying capacity restrictions that directly affected the size of the group that sailed in them, an idea that aligns well with Bjerck's assumption (2016).

The incorporation of the *dalka* into the western channels and the Strait of Magellan would have modified some traditional patterns regarding what was observed in the bark canoes. Longer hulls and more displacement allowed the transport of a greater number of people on board (Figure 6.6). Results indicate plank canoes with 8, 9, and even 15 adults. This type of canoe would have allowed for the integration of two or three families in the same permanent crew, reducing the need for two canoes to operate simultaneously and incorporating new forms of organization on board (Aguilera 2023). Moreover, more adult crew should have led to more paddling power, with greater travel speeds. Nev-

Table 6.1. Number of Men, Women, and Children Observed in a
Canoe in the Study Area

	Bark canoe	Dalka	Dugout
NUMBER OF CASES	41	23	7
MAXIMUM	7* (11)	15	5 (12.3)
MINIMUM	2	1	2 (4.3)
AVERAGE	4.1 (5.9)	5.7 (7.7)	3.8 (7.8)

*All numbers outside the parenthesis refer to adults; numbers in parentheses
include children.

ertheless, there are also observations of *dalkas* with 3–4 adult paddlers, like
in the bark canoe, suggesting that even when the limiting physical factor was
removed, social customs regarding crew size remained in place for some cases.

This interpretation does not apply to the incorporation of dugout canoes at
the end of the nineteenth century since this type of hull did not significantly
increase the carrying capacity (in terms of functional space available per person)
with respect to the bark canoes (Figure 6.6). However, the social effects of the
adoption of this boat would have had other manifestations, especially in terms
of the forms of operation. It is important to note that the adoption of dugout
canoes occurred at the time of demographic collapse of the canoe communities of the Fuegian archipelago and the Strait of Magellan, so the population
(and, thus, the crews) was a fraction of that reported at the time of first contact
(Emperaire 2014; Orquera and Piana 1999). The increase in the number of
children observed in the same canoe (Table 6.1) may suggest a change in the
reproductive behavior of Indigenous people to counteract diseases, kidnap-
pings, and other social and demographic threats.

Conclusions

The canoe was the vehicle through which canoe communities took advantage
of the geography of the Patagonian channels. It was the platform that allowed
access to resources that enabled them to meet the challenges of survival and
thrive in a marine landscape. It was the center of cultural construction and
social meanings for the phenomena of the universe and, equally, it was a means
of resistance and escape from the expansion of the new occupants of their
traditional spaces. The canoe encloses a world of knowledge about available
resources and material engineering. Indigenous navigation holds the keys to
understanding a differentiated attribute of the communities that used them—
that is, the relationship with the world from the sea.

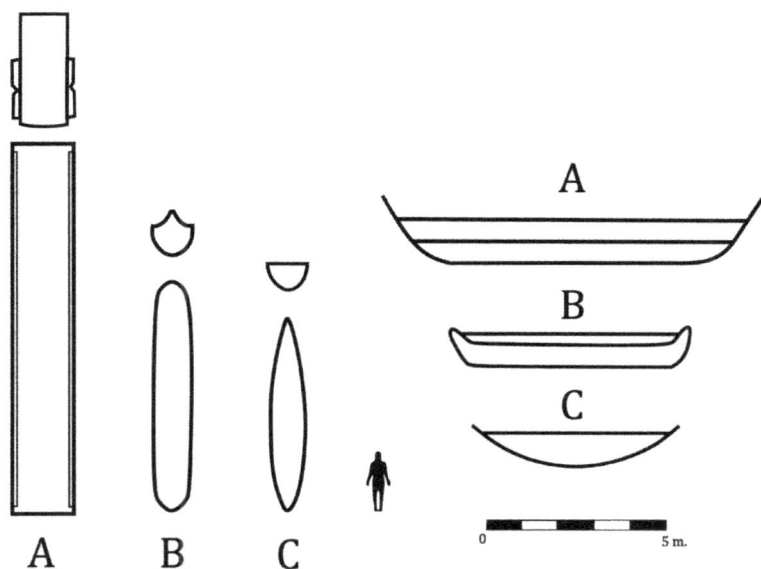

Figure 6.6. Plan view, longitudinal and transverse section of the three types of Patagonian canoes analyzed in the chapter. Length, beam, and draft measures are based on the average measures of the database. Illustration by Javiera Aguilera.

The adoption of new forms of construction and navigation techniques among the canoe peoples of southern Patagonia and Tierra del Fuego responded to social processes broader than cultural diffusion, suggesting dynamic processes of change. At the moment of first contact, the bark canoe was the vernacular watercraft from the southern Gulf of Penas to Cape Horn. Its distribution and descriptions confirm that it was a stable technological solution over time, as it continued to be used until the twentieth century in some areas, particularly around the Beagle Channel by Yagán communities. In the western channels and the Strait of Magellan, homeland of the Kawésqar, the bark canoe was rapidly replaced by plank canoes after the end of the eighteenth century. The chronology and direction of the *dalka* spread leads us to suggest that this process was originated by the southward migration of Chonos communities after 1740. A third Indigenous watercraft innovation is the spread of the dugout canoe in Fuego-Patagonia. This was the result of two independent focuses of technological adoption that happened in the two extremes of the region. Notably, this type of canoe, the last type of Indigenous watercraft described in the area, is known as the traditional one for the Kawésqar community of Puerto Edén in the late 1950s (Emperaire 2014).

Indigenous canoe populations generated or incorporated technological innovations at an outstanding speed. Based on the analysis of sightings, the replacement of traditional forms by new technological solutions occurred within 50 years, virtually two generations, which implied the total abandonment of practices in exchange for new solutions. This versatility suggests that the forms recognized as traditional correspond to the last incorporation adopted by a community. Just one century after the adoption of the dugout canoe in the channels of the Fuegian archipelago, ethnographer Gusinde could not find anyone who knew how to craft a bark canoe, regardless of the past strength of the "human–boat relation" (Bjerck 2016). The dugout canoe became the traditional canoe, the Indigenous one. All this happened after the social and demographic collapse caused by missionary and industrial activity, which could explain the depth of this process.

However, a few decades earlier, the replacement of the bark canoe by the *dalka* also was a relatively fast process, although spatially more limited to the western channels. A better understanding of the reasons behind this change, particularly of the nautical advantages of this new type of watercraft, could be enlightening. Contrary to previous thought, the use of *dalka* was not restricted spatially or built exclusively from forest species like the Guaitecas cypress or the larch. The raw materials used in the elaboration of the planks were diverse and did not adjust with the main ethnographic descriptions, evidencing great versatility and flexibility in the extraction of wood planks. Furthermore, even if this watercraft could be related to the apparition of polished stone technology in northern Patagonia, in southern Patagonia it was built with the same toolkit that had been used previously for crafting the bark canoe (Aguilera 2023). All this suggests a very dynamic nautical history during the last centuries that must caution us about the direct use of ethnographic analogies in the analysis of the archaeological record. Moreover, further analysis of the social changes derived from the adoption of new watercraft like the plank canoe could have larger implications in the study of how "human–boat relations" were built in the past, both in Patagonia and in any other watery world.

References Cited

Aguilera, Nelson. 2023. Aqa K'énak: Navegación Indígena en Patagonia 1520–1960: Una aproximación desde la etnohistoria y la arqueología. PhD dissertation, Departamento de Arqueología, Universidad Autónoma de Barcelona, Bellaterra.
Aguilera, Nelson, Alberto García-Piquer, Raquel Piqué, and Alfredo Prieto. 2021. Fragmentos de realidad: Arqueología y etnografía de las canoas de corteza en el área del Cabo de Hornos. *Anales del Museo de América* 27: 6–34.

Aldunate, Carlos. 1996. Mapuche: Gente de la tierra. In *Etnografía: Sociedades indígenas contemporáneas y su ideología,* edited by Jorge Hidalgo L., Virgilio Schiappacasse F., Hans Niemeyer F., Carlos Aldunate del S., and Pedro Mege R., pp. 111–134. Editorial Andrés Bello, Santiago.

Ames, Kenneth. 2002. Going by Boat: The Forager-Collector Continuum at Sea. In *Beyond Foraging and Collecting: Evolutionary Change in Hunter-Gatherer Settlement Systems,* edited by Ben Fitzhugh and Junko Habu, 19–52. Kluwer Academic / Plenum Publishers, New York.

Arnold, Béat. 2017. *Bark Canoes in South America from Amazonia to Tierra del Fuego.* Editions D'Encre, Le Locle.

Bertrand, Alejandro. 1886. Memoria sobre la rejion central de las tierras Magallanicas. In *Anuario Hidrográfico de la Marina de Chile,* Vol. 11, pp. 203–343. Imprenta Nacional, Santiago.

Bird, Harry. 2012. *Through Her Eyes: Adventures of Margaret McKelvy Bird.* Lulu Press.

Bird, Junius. 1988. *Travels and Archaeology in South Chile.* University of Iowa Press, Iowa City.

Bjerck, Hein. 2016. Settlements and Seafaring: Reflections on the Integration of Boats and Settlements among Marine Foragers in Early Mesolithic Norway and the Yámana of Tierra del Fuego. *Journal of Island and Coastal Archaeology* 12(2): 276–299.

Bridges, Lucas. 1952. *El Último Confín de la Tierra.* EMECE Editores, Buenos Aires.

Bridges, Thomas. 1930. *A Short Account of Tierra del Fuego and Its Inhabitants by Thomas Bridges (writings compiled by his children, 1930).* Electronic document, https://patlibros.org/tdf/doc.php, accessed June 27, 2024.

Bukeley, John, and John Cummins. 1743. *A Voyage to the South-Seas, in the Years 1740–1: Containing, a Faithful Narrative of the Loss of His Majesty's Ship the Wager on a Desolate Island in the Latitude 47 South, Longitude 81:40 West.* J. Robinson, London.

Burg, Amos. 1937. Inside Cape Horn. *National Geographic Magazine* 72(6): 343–387.

Byron, John. 2017. *La Pérdida de la Fragata Wager: Naufragio y supervivencia en la Patagonia.* Ediciones Universidad Diego Portales, Santiago.

Carabias, Diego. 2019. Canoas monóxilas etnográficas de los nómadas canoeros de la Patagonia Occidental y Tierra del Fuego del Museo de Historia Natural de Valparaíso. *Rastros en el agua: Exploradores, embarcaciones y materialidades,* edited by María Dolz, pp. 105–132. Ediciones Servicio Nacional del Patrimonio Cultural, Santiago.

Darwin, Charles. 1840. *Journal of Researches into the Geology and Natural History of the Various Countries Visited by H.M.S. Beagle, Under the Command of Captain Fitz-Roy from 1832 to 1836 by Charles Darwin.* Henry Colburn, London.

Emperaire, Joseph. 2014. *Los Nómades del Mar.* LOM Ediciones. Santiago.

Estay, Sergio, Daniela López, Carmen Silva, Eugenia Gayó, Virginia Mcrostie, and Mauricio Lima. 2022. A Modeling Approach to Estimate the Historical Population Size of the Patagonian Kawésqar People. *The Holocene* 32(6): 095968362210807.

Finsterbush, Carlos. 1934. Las dalcas de Chiloé y los chilotes. *Revista Chilena de Historia y Geografía* 75(82): 412–433.

Fitz Roy, Robert. 1839. *Narrative of the Surveying Voyages of His Majesty's Ships Adventure and Beagle Between the Years 1826 and 1836, Describing Their Examination of the Southern Shores of South America, and the Beagle's Circumnavigation of the Globe.* Henry Colburn, London.

García, José. 1889. Diario del viaje i navegacion hechos por el padre Jose Garcia de la Compañía de Jesus desde su mision de Cailin, en Chiloe, hacia el sur en los años 1766 i 1767. *Anuario Hidrográfico de la Marina de Chile,* Vol. 14, pp. 3–47. Imprenta Nacional, Santiago.

Gusinde, Martin. 1986. *Los Indios de la Tierra del Fuego, Los Yámana.* Centro Argentino de Etnología Americana, Consejo Nacional de Investigaciones Científicas, Buenos Aires.

Hawkesworth, John. 1773. *An Account of the Voyages Undertaken by the Order of His Present Majesty for Making Discoveries in the Southern Hemisphere, . . . By John Hawkesworth, LL.D. in Three Volumes,* Vol. 2. W. Strahan and T. Cadell, London.

Hudson, Francisco. 1857. Reconocimiento en los Canales del Sud de Chiloé, ejecutado en el Janequeo. *Memoria que el Ministro de Estado en el Departamento de Marina presenta al Congreso Nacional 1857,* pp. 58–66. Imprenta del Ferrocarril, Santiago.

Kent, Rockwell. 1924. *Voyaging Southward from the Strait of Magellan.* Halcyon House, New York.

Latcham, Ricardo. 1930. La dalca de Chiloé y los canales patagónicos. *Boletín del Museo Nacional de Chile* 13: 63–72.

Legoupil, Dominique, and Denis Chevallay. 2017. La Terre du Feu (1925), Un documental francés sobre los indígenas de Patagonia y Tierra del Fuego. *Magallania* 45(2): 67–80.

Lemay, Gaston. 1881. *À Bord de "la Junon."* G. Charpentier Éditeur, Paris.

Lira, Nicolás. 2015. Embarcations de la tradition indigène en Patagonie du nord/sud Chili: Typologie, technologie et routes de navigation de la cordillère des Andes à la mer. PhD dissertation, Université de Paris I, Pantheón—Sorbonne, Paris.

Lira, Nicolás. 2018. Embarcaciones de tradición indígena en Patagonia septentrional: Arqueología, Historia y Etnografía. *Revista de Arqueología Histórica Argentina y Latinoamericana* 12(1): 7–36.

Lira, Nicolás, Valentina Figueroa, and Romina Braicovich. 2015. Informe sobre los restos de la dalca del Museo Etnográfico de Achao, Chiloé. *Magallania* 43(1): 309–320.

Lothrop, Samuel K. 1932. Aboriginal Navigation Off the West Coast of South America. *Journal of the Royal Anthropological Institute of Great Britain and Ireland* 62: 229–256.

Lozano, Pedro. 1755. *Historia de la Compañía de Jesús de la provincia del Paraguay,* Vol. 2. Imprenta de la Viuda de Manuel Fernández, Madrid.

Mac-Érin, U. 1897. *Huit mois sur les deux océans: Voyage d'études et d'agrément.* Alfred Cattier Éditeur, Paris.

Maldonado, Roberto. 1915. Memoria sobre los trabajos hidrográficos realizados en los Archipiélagos Patagónicos en 1912. *Anuario Hidrográfico de la Marina de Chile,* Vol. 29, pp. 86–232. Imprenta de la Armada, Valparaíso.

Martinic, Mateo. 1989. Los canoeros de la Patagonia meridional: Población histórica y distribución geográfica (siglos XIX y XX). El fin de una etnia. *Journal de la Société Des Américanistes* 75: 35–61.

Medina, Alberto. 1984. Embarcaciones chilenas precolombinas, La Dalca de Chiloé.

Revista Chilena de Antropología (4): 121–138. Facultad de Filosofía, Humanidades y Educación, Universidad de Chile, Santiago.

Medina, José. 1923. *Memorias de un Oficial de Marina Inglés al Servicio de Chile Durante los Años de 1821–1829.* Imprenta Universitaria, Santiago.

Moraleda, José. 1796. *Diario de la navegación desde el puerto del Callao de Lima al de San Carlos de Chiloé y de este al reconocimiento del archipiélago de Chono y costa occidental patagónica comprendida entre los 41 y 46 grados de latitud meridional, hecho de real orden y comisión del Exmo. Sr. Francisco Gil, virrey del Perú, por José de Moraleda y Montero, alférez de fragata y primer piloto de la Real Armada.* Manuscript on file, Fondo Dirección de Hidrografía, Sección Expediciones, Biblioteca Virtual de Defensa, Madrid.

Orquera, Luis, and Ernesto Piana. 1999. *La Vida Material y Social de los Yámana.* Editorial Universitaria de Buenos Aires, Buenos Aires.

Pallo, María. 2011. Condicionamientos de la dinámica ambiental en las decisiones humanas sobre asentamiento y circulación a lo largo del estrecho de Magallanes durante el Holoceno Tardío. *Magallania* 39(2): 177–192.

Prieto, Alfredo, Denis Chevallay, and David Ovando. 2000. Los pasos de indios en Patagonia Austral. In *Desde el país de los gigantes. Perspectivas arqueológicas en Patagonia,* Vol. 2, edited by Juan Bautista Belardi, Flavia Carballo Marina, and Silvana Espinosa, pp. 87–94. Universidad Nacional de la Patagonia Austral, Río Gallegos, Argentina.

Puente, Manuel. 1986. La dalka de Chiloé y su influencia en la exploración austral: Contribución a su estudio. *Revista de Historia Naval* 4(15): 19–44.

Reynaud, Pierre. 1876. Rapport médical sur la campagne de l'aviso Le Hermitte. *Archives de Médecine Navale,* Vol. 26, pp. 81–104. Librairie J.B. Bailliere et Fils, Paris.

Rivas, Pilar, Carlos Ocampo, and Eugenio Aspillaga. 1999. Poblamiento temprano en los canales patagónicos, núcleo ecotonal septentrional. *Anales del Instituto de la Patagonia* 27: 221–230.

Roig, Fidel, and José Boninsegna. 1991. Estudios sobre el crecimiento radial, basal, en altura y de las condiciones climáticas que afectan el desarrollo de Pilgerodendron uviferum. *Revista Chilena de Historia Natural* 64: 53–63.

Rudmose, Robert, Robert Mossman, and James Harey. 1906. *The Voyage of the "Scotia" Being the Record of a Voyage of Exploration in Antarctic Seas.* W. Blackwood and Sons, Edinburgh and London.

Techo, Nicolas del. 1897 [1673]. *Historia de la provincia del Paraguay de la Compañía de Jesús,* Vol. 2, Book 3, Chapter 8. Librería y Casa Editorial A. de Uribe y Compañía, Madrid, Asunción del Paraguay.

Thiers, Oscar, and Victor Gerding. 2007. Variabilidad topográfica y edáfica en bosques de Nothofagus betuloides (Mirb) Blume, en el suroeste de Tierra del Fuego, Chile. *Revista Chilena de Historia Natural* 80(2): 201–211.

Urbina, Ximena. 2017. Traslados de los indígenas de los archipiélagos patagónicos occidentales de Chiloé en los Siglos XVI, XVII y XVIII. In *América en Diásporas, Esclavitudes y Migraciones Forzadas en Chile y Otras Regiones Americanas,* edited by Jaime Valenzuela, pp. 381–411. RIL Editores, Instituto de Historia, Pontificia Universidad Católica de Chile, Santiago.

Urbina, Ximena. 2018. *Fuentes para la historia de Patagonia occidental en el periodo co-*

lonial. Segunda Parte, Siglos XVIII. Ediciones Universitarias de Valparaíso, Pontificia Universidad Católica de Valparaíso, Valparaíso.

Vidal, Francisco. 1880. Documentos relativos a la historia náutica de Chile. *Anuario Hidrográfico de la Marina de Chile,* Vol. 6, pp. 428–564. Imprenta Nacional, Santiago.

7

Collective Action, Transport Costs, Watercraft Technologies, and the Engineered Ancestral Landscapes of Southern Florida

Watercraft technologies have a long history in southern Florida. Archaeologists have recovered large vessels, but historic documents also describe the Calusa as using complex vessels able to transport large numbers of people. In addition to the sizable amount of labor that the people of the region invested in building such watercraft, they constructed canal systems throughout southern Florida. These canals not only facilitated movement within the bays and estuaries but also connected the coast with a vast network of settlements in the region's immense interior river and wetlands systems. These canal networks required labor to build as well as considerable maintenance to keep them free and clear for navigation. Further, the incorporation of navigable rivers into this system presented additional challenges and required detailed knowledge of hydrological engineering.

Here I consider the nature of watercraft technologies and canal systems in the context of collective and cooperative institutions (Blanton and Fargher 2016; see also Fauvelle et al., this volume: Chapter 1). Specifically, I consider the social and political rules under which these technologies and institutions emerged. My argument considers the idea that most of the large-scale canals in the region did not require centralized hierarchical leadership. Such canals did, however, pose several different collective action challenges not only in their creation but also their maintenance. The histories of canal construction were likely also related to the social relations of watercraft production. I consider the idea that the solidification of social ties and institutions (e.g., Holland-Lulewicz et al. 2020) that canals created among participating communities and the histories of watercraft technologies engendered constellations of technologies and engineered landscapes that formed a kind of armature for the emergence of regional political systems of the kind observed in the area in the sixteenth century.

To explore the role that canal construction and watercraft technologies played in Calusa society and politics, I first provide a broader discussion of watercraft technologies in southwest and greater Florida. Next I discuss the rise of the Calusa polity as we currently understand its formation. Following this I summarize the interpretations of canals at Calusa and related settlements—specifically considering the idea that canals were for moving tribute. Finally, to put canals in a broader context, I explore the history of their construction in the region along with changes in the diversity and capacities of watercraft over time. In the concluding section I bring this discussion together to explore how these histories contribute to our overall understanding of how the Calusa and likely other Indigenous peoples of the region solved unique challenges in both social and technological ways (see also Grier et al., this volume: Chapter 13).

Watercraft Technologies

Before considering how watercraft technologies and broader landscape modifications articulated with Calusa histories, I first describe our broader understanding of watercraft in greater Florida. While the use of watercrafts likely existed in the Pleistocene, the earliest evidence we have in Florida dates to around 7,000 years ago (Wheeler et al. 2003). Archaeologists have recovered canoes around the state, with the largest number of canoes found in a single location being over 100 canoes at Newnans Lake dating between 2,300 and 5,000 years ago (Wheeler et al. 2003).

Both the documentary and archaeological records speak to the wide variety of watercraft constructed in Florida. The large sample from Newnans Lake indicates that size does not correlate with the age of the canoe, and the average size of these early ones is on the order of around 5.6 m long, with at least one example of 9.5 m (Wheeler et al. 2003: 545). The vast majority of the Newnans canoes are made of pine (*Pinus* spp.) with a few unidentified hardwoods (Gymnospermae) and one bald cypress (*Taxodium distichum*) (Wheeler et al. 2003: Table 1). As Ryan Wheeler and colleagues (2003: 542) note, previous studies suggest that pine may be the favored material during the earlier ancestral period in Florida, but certainly early builders did not eschew cypress as one of the oldest canoes was constructed of this material.

Early documents detailing the Calusa in southwest Florida also provide information on vessel size and their ubiquitous presence on the landscape.

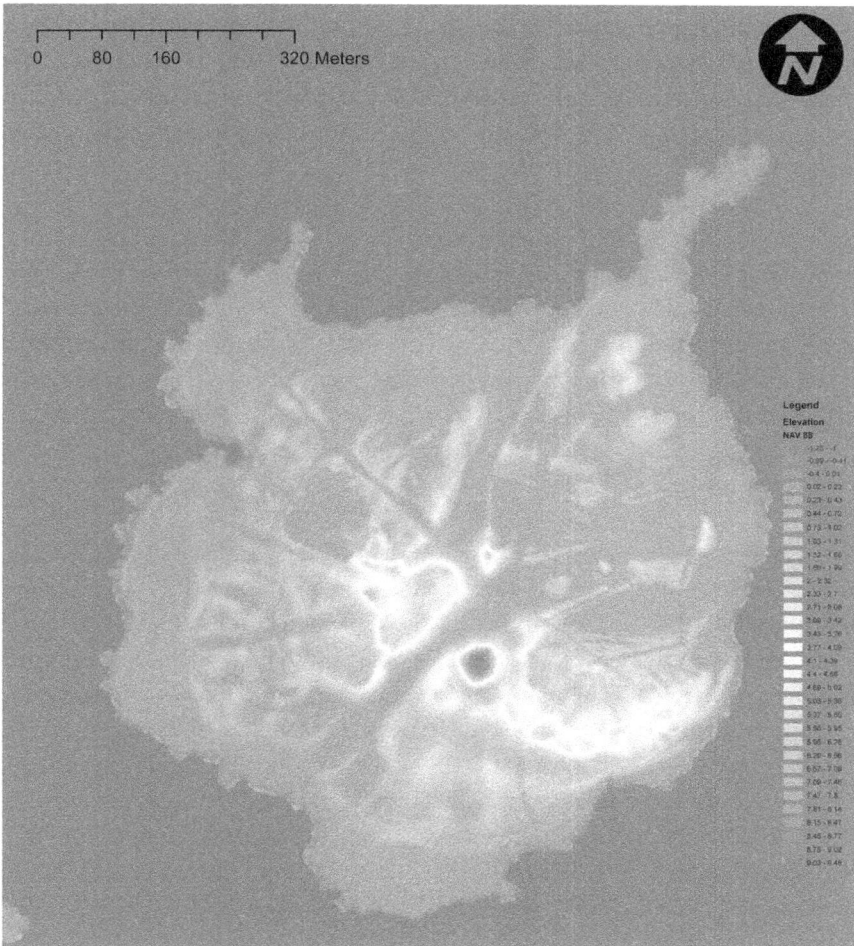

Figure 7.1. Lidar of the Mound Key site showing the main canal and its subsidiary canal system.

There are several descriptions of many canoes being mustered during a short period. This is particularly the case for some of the descriptions of armed conflict. For example, one account during 1612 describes 60 war canoes sent to Tampa Bay to engage the Spaniards there (Hann 1991: 10), and an earlier account describes "eighty shielded canoes" attacking Ponce de León's crew (Worth 2014: 18). Another 1614 account indicates a force of over 300 canoes sent to Mocoço (a province) along Tampa Bay to attack the towns allied with the Spaniards (Thompson, Roberts Thompson et al. 2018).

These descriptions indicate they were not small vessels but were substantial, often carrying large numbers of people. Some of these vessels were elaborate and were tandem canoes lashed together (Worth 2014: 52). Perhaps one of the most elaborate descriptions is that of the vessel that carried Callus, which had two canoes "tied together and covered and outfitted very well with awnings of arches and mats" (Worth 2014: 256).

From my brief discussion here, it is clear that both the size and structure of watercraft varied through time; however, it is not a simple linear relationship. In fact, it seems that watercraft technologies and the scale at which vessels were constructed were in place for thousands of years prior. So what role did these watercrafts play in the organization of the Calusa during the sixteenth century? To understand this, we first need to delve into the different perspectives regarding the trajectories of Calusa political organization.

The Calusa Polity

In 1566 Pedro Menéndez de Avilés arrived at the cultural site of Mound Key, which we are now certain was the capital of the Calusa polity (Thompson, Marquardt et al. 2018; Thompson, Roberts Thompson et al. 2018) (Figure 7.1). More properly referred to as Caalus or Calos, the settlement was the apex of a polity that stretched from Charlotte Harbor to the Florida Keys and included communities that lived in the interior around modern Lake Okeechobee (Thompson, Marquardt et al. 2018). In all, the polity encompassed some 50 to 60 communities and at least 20,000 people (Thompson and Worth 2011; Worth 2014). The coastal settlements were densely populated, and Mound Key had a population of an estimated 4,000 people and exhibits many of the traits and processes that archaeologists working in other areas of the world consider urban (Thompson 2023b).

Scholars have labeled the sixteenth-century Calusa polity as a weak tributary state (Marquardt 1987, 1988, 2014) or a complex chiefdom (Widmer 1988), or they have chosen to eschew typologies in favor of a view that emphasizes processes and institutions (Thompson, Marquardt et al. 2018). Regardless of the viewpoint, by the sixteenth century the Calusa capital held political sway over a vast region, a point that few scholars debate. That said, exactly when this large polity emerged is difficult to pinpoint. Randolph Widmer (1988: 272–273, 279) argues that the Calusa were a "highly developed chiefdom" that exhibited "no evolutionary change from AD 800 until contact." This complexity, he argues, was largely attributable to a stabilization of sea level and a rising population reaching its carrying capacity (Widmer 1988: 222–223, 229–255). William Marquardt (2014), on the other hand, while agreeing that some degree of political centralization occurred around AD 800 (i.e., simple chiefdoms), posited that the political histories of the region were more variable and influenced to some degree by large-scale shifts in climate and local sea levels, limiting the availability and distribution of species in the estuaries variously over time. Others suggest that the Calusa political complex as it existed in the sixteenth century emerged relatively recently in their history.

Marquardt (1988) also favors a view that the sixteenth-century Calusa political formations were likely the result of an influx of Spanish goods and as such were a post-contact-period phenomenon. Similarly, others have also suggested that the sixteenth century of Calusa political formulation was a relatively recent occurrence either right at or just before the beginning of the 1500s (Thompson, Marquardt et al. 2018). While these various viewpoints offer some evidence to these shifts in political organization, we currently do not possess the types of data more commonly used to assess political centralization (e.g., settlement/ survey data indicated site hierarchy). In fact, there are many larger (ca. 20 ha) settlements occupied for extended time frames located at relatively close distances to one another; therefore, the traditional assessments of this kind might not be applicable, and it is possible that Calusa political integration is dissimilar to the inland setting, where archaeologists developed most of these models.

The research at Mound Key offers probably the best current insight into regional political integration. Excavations there revealed a series of construction stages of a large structure on the summit of Mound 1. The last identifiable construction dates to the sixteenth century, the "king's house," as referred to in the accounts of Menéndez de Avilés' arrival at the capital, where it is described as being able to hold around 2,000 people—a figure substantiated to a large degree by the number of people in attendance at Menéndez's reception. Additional work and coring allow us to connect these construction stages to a series of other large-scale projects at the settlement. The Calusa constructed the earliest "king's house" around cal. AD 1000, and at that time the two largest mounds (Mound 2 and 1) were almost as tall, ca. 6 m and 10 m, respectively, as they were in the sixteenth century. The settlement's inhabitants constructed the main canal that bisects the settlement around this time as well. The second rebuilding of the Mound 1 structure occurred between cal. 1300 and 1400, coinciding with the building of two large water courts for large-scale live surplus storage of fish at the southern end of the settlement (Thompson et al. 2020). The final stage of construction for the Mound 1 structure that is archaeologically resolvable occurred just before or during the early part of the sixteenth century. This last phase of construction is definitively associated with the existence of the large-scale Calusa polity described by the Spaniards who observed it firsthand.

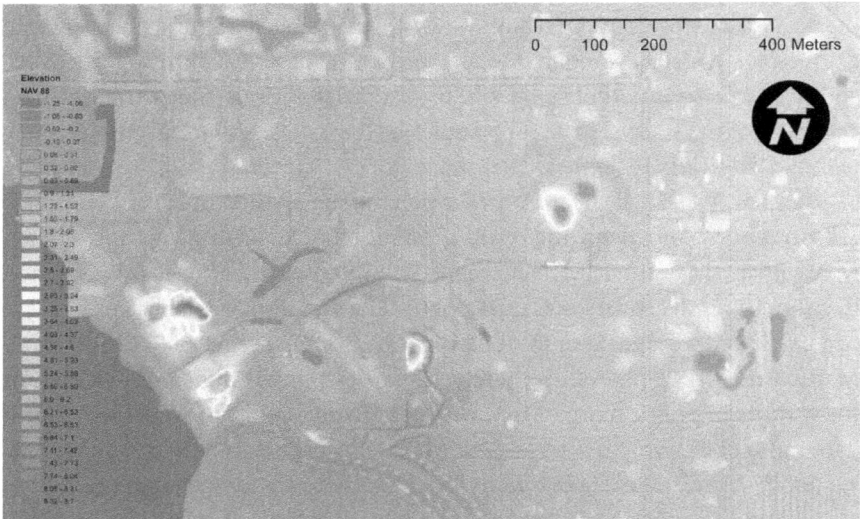

Figure 7.2. Lidar of the Pineland site showing the main canal and the large mounds adjacent to it.

What we know from this work is that the structure on Mound 1 at Mound Key was the seat of authority for the polity. Were the other earlier structures evidence of this authority that extends back in time some 500 to 700 years up until the abandonment of the settlement in the 1700s? Certainly, this is possible, and researchers argue that the rebuilding of the structure is evidence of a long-lived powerful corporate group (Thompson, Marquardt et al. 2018). However, that the settlement held sway over the vast territory it exacted tribute from in the sixteenth century cannot be substantiated by this evidence alone. And there is evidence of many other large and likely powerful, in a political sense, settlements along the coast of the Calusa heartland. These settlements, too, had large-scale labor projects and massive shell architecture.

Pineland is a good example of one of these large-scale neighboring settlements (Figure 7.2). Like Mound Key, the settlement has a long occupational history (e.g., ca. cal. AD 50–1700s) and large shell architecture in the form of linear platform mounds bisected by a large canal with a series of subsidiary canals and water impoundments. Unlike Mound Key, Pineland is not an anthropogenic island but rather sits at the water's edge of a back barrier island within Charlotte Harbor.

As with Pineland and Mound Key, many Calusa settlements have a canal that bisects them, or they are in proximity to one. What role did these "roadways" play in Calusa society? What are their larger histories? To begin to explore answers to these questions, we need to first resolve the larger temporality of canal systems in Florida and where they occur in the region.

The Scale and Temporality of Canal Systems and Water Courses

Apart from Alaska, Florida has the longest coastline of any state and has over 1,700 waterways. It is a landscape where water is the defining feature, especially in the low-lying areas of southern peninsular Florida. It is no wonder, then, that the region's Indigenous people have used these waterways for travel for millennia.

In one of the most important studies on the subject of canoe travel in Florida, Julie Duggins (2019) argues that site distributions, geology, and the distribution of canoes from many time periods recovered at the edges of watersheds indicate a large complex system with deep time. One of the key facets of Duggins' "Hydro-Highway Hypothesis" is that canoes are cached at transit points between waterways, and there are even historic documents that describe the process of canoes caching among the Seminoles (Duggins 2019: 94). Most recently, I suggested the possibility that cached canoes could possibly represent a common pool system and that perhaps many people could use the cached canoes (Thompson 2023a).

Along with Florida's considerable navigable waterways, the people of the region also constructed canal systems for canoe transport. Some of these canals connected various settlements. Other times they connected settlements with other navigable waterways. Most of these canal systems are in the southern part of Florida, but a number of them are found in the northern part of Florida and along the Gulf Coast of Alabama (Luer 1989; Rodning 2003; Waselkov et al. 2022; Wheeler 1995, 1998).

The size of canals in the region varies. It is possible that some canals reached lengths of over 15 km and others were as short as 1.6 km; perhaps 4 km is a good average for the range (Thompson 2016: 330). This average distance may relate to the distance between earthworks, which is around 5 km (Thompson 2016). Critical to understanding the nature of these large-scale, labor-intensive constructions is their timing. As Wheeler et al. (2003: 548) note, "attempts to date canals have met with difficulty." However, archaeologists now have secured dates for several canals, including the Ortona canals in the Lake Okeechobee and Caloosahatchee River vicinity, dating them between AD 110–380 and AD 600–890 (Carr et al. 2002: 16, 21; Wheeler et al. 2003). Gregory Waselkov et al. (2022) dated an approximately 1.39-km-long canal located along the Gulf Coast of Alabama to around AD 600. The critical study by Waselkov and colleagues also explores the functioning of canals and builds upon the work of George Luer and Wheeler (1997: 122–124), who maintain that such structures were not merely ditches with water in them but also were feats of hydroengineering that required check dams at inflection points so they could maintain a certain water level (Waselkov et al. 2022: 495–496) (Figure 7.3). From the small sample of dated canals, we can infer that they were in use between AD 500 and AD 1500 over vast areas of Florida.

The Pine Island canal is perhaps the most intensively studied Florida canal. The main canal bisects the settlement but also extends far inland and traverses the entirety of Pine Island to Matlacha Channel (Marquardt and Walker 2013). Stretching some 4 km with a possible extension into the interior mainland, the Pine Island canal was a massive undertaking. In the current preserved sections of the canal, it varies from 5.5 m to 7.1 m wide and is over 1 m deep (Luer and Wheeler 1997). The canal is thought to have been constructed during the Caloosahatchee IIB period (ca. AD 800 to 1200) (Marquardt 2013: 13; Marquardt and Walker 2013), although no definitive radiocarbon ages have been run for the feature.

The Pine Island canal integrated features that had long histories and in general tied together a whole wetland landscape at the settlement's core. Recent remote sensing, coupled with a long-term excavation program at the settlement, reveals a vast landscape with impoundments and subsidiary canals that offer access to various parts of Pineland, including its ancestral burial mound, which were surrounded by impoundments and canals (Thompson et al. 2014). Thus, canals facilitated not only movement between settlements but also intrasettlement travel as well (Thompson 2023b).

This discussion illustrates that the Indigenous peoples of Florida had in place a vast network that allowed for canoe travel that was both a natural system of rivers and a culturally constructed landscape through the building of canals and caching of canoes. This allowed people over thousands of years to traverse the vast interior wetlands of Florida, to travel direct to other waterways and settlements on constructed canals during later periods, and even to move about within settlements with ease. Even the Spaniards noted the ease of travel between sites, as Lopez de Valasco wrote about travel from Mound Key through Matanzas Pass and across the waters of San Carlos Bay to get to the settlement of Tampa (i.e., Pineland) in Pine Island Sound (Worth 2013: 768). The final piece of our consideration of transport is long-distance travel over open ocean water.

Although not directly stated or engaged by scholars to any large degree, the peoples of southern Florida, and the Calusa specifically, traversed the open waters of the Gulf of Mexico in their watercraft. Several different references attest to this fact in Spanish documents. First, we do know that south Florida

Figure 7.3. Artist reconstruction of the Pine Island canal, a 4-km-long canal that bisects the island. Image Courtesy of the Florida Museum of Natural History.

groups fished or hunted from canoes in open water. Specifically, there is reference to whaling, and the documents indicate that they go out to sea in canoes to do this (Hann 1991: 312). And while the description of hunting is rather odd, archaeologists have recovered numerous whale bones at sites in southern Florida, especially among the Tequesta (Carr 2012). Yet another instance tells of travel by canoe from St. Augustine via the "River of St. Martin" to the "bay of Apalachee," then on to "Carlos [Mound Key], skirting the coast," a trip that covered hundreds of kilometers (Hann 1991: 25). Travel from the Florida Keys from Mound Key is frequently described in the documents as well, which would have required open-water travel (see Hann 1991: 41, 198–199). Indeed, travel between the islands of the Florida Keys also required open-water travel.

Up to this point I have described travel along the coast or in sight of it. Could travel between Florida and the Greater Antilles have occurred? This is likely, as one documented instance points to this possibility. John Hann (1991: 330) describes several Native peoples following a Spanish vessel to Cuba, arriving only four days after it, suggesting that the Natives traveled via canoe. That south Florida Native people were capable of open-water travel over great distances should not be a surprise; after all, the seafaring abilities and technologies in this region are no less than of those groups that colonized the islands of the Greater and Lesser Antilles (see Fitzpatrick 2013; Shearn 2020).

Some archaeologists might take exception to the possibility of such connections, citing the lack of pottery similarity or the lack of the adoption of cultivars, such as manioc and maize (see Knight and Worth 2021 for a discussion). While these are valid points, the Calusa and other south Florida groups were in contact with northern Florida people and did not adopt their crops (i.e., maize) or ceramics. So this lack of evidence is not definitive. Other evidence for connections is the presence of chili pepper and papaya at Calusa settlements (i.e., Pineland) (Newsom and Scarry 2013) and the fact that when Ponce de León arrived, there were already Spanish-speaking Native people from the Caribbean living among the Calusa (see Worth 2014) and even a settlement within the Calusa domain of Cuban Indian refugees (Worth 2014: 17).

Transport Costs, Sunk-Cost Effects, and Public, Private, and Common Pool Resources

The traditions and institutions of the Calusa and the other groups of southern Florida depart dramatically from those of the interior riverine American southeast. One of the key differences is that many of the settlements, especially those located along the coastal margins, have long histories, but this is also true of many of the inland sites. For example, both Mound Key and Pineland evidence persistent occupation for over 1000 years (Marquardt et al. 2022; Marquardt and Walker 2013; Thompson et al. 2016). Similarly, inhabitants of interior settlements along the Lake Okeechobee basin also were constructing and occupying specific locales from ca. 800 BC to the 1700s (Thompson and Pluckhahn 2012). These are by no means the only examples of such long-lived settlements, and the answer to why this is the case can in part be explained by two specific phenomena: transport costs and sunk-cost effects.

Given the waterways and canal systems described above, compared to the inland more northern southeast, transport costs were quite low comparatively. Robert Drennan's (1984) evaluation of transport costs in Mesoamerica provides some reasonable figures for understanding this in southern Florida. He notes that water transport is highly efficient and that canoes in both Mesoamerica and South America ranged up to 10 m in length and not only could carry a considerable number of people but also had displacement loads of up to and likely over one metric ton (Drennan 1984: 105–106). These figures seem reasonable for Florida canoes as well, which are around the same general size class as the ones Drennan described. Even loaded down, canoes in Mesoamerica could travel considerable distances, up to 40 km downstream and 20 km a day paddling against a current (Drennan 1984: 106). Again, given what we know about the distances traveled during the period that the Spanish recorded such trips, this seems to be a good estimate (see the discussion above).

From the foregoing, we can infer that transport costs in Florida were much lower than in other regions of the American southeast. This has two implications. The first is that the barriers to the transport of needed goods to a given settlement from another settlement or region were relatively easy to overcome. If needed resources can be transported and harvested at larger distances from a given settlement, then there possibly was less need to frequently relocate settlements. There may have been other reasons inhabitants would have invested in maintaining sustainable settlements, such as the investment in infrastructure.

Previously I argued that the investment in infrastructure—specifically, the construction of canals—was a kind of sunk-cost effect for settlements (Thompson 2016). When settlements make the decision to build a canal, this not only expends a lot of labor, it also creates a pathway and link between people who live along the canal or who have easy access to it. Canals need to be maintained so that infilling with plants or sediments does not render them unusable. Thus, the construction of canals sets up the need to solve collective action problems that would not be one-time events but would require rules, regulations, and ways to deal with any social issues that may arise (e.g., who cleans and maintains canals). In other words, people who engaged in canal construction would need to deal with any issues that would be associated with such communal resources (e.g., noncooperators) (Blanton and Fargher 2016: 30).

As Richard Blanton and Lane Fargher (2016) note, collective action solutions are often interwoven with institutions. Following this line of thinking, I argue that canal construction would engender builders and users of such facilities to solve collective action problems, and such solutions would likely be regularized within broader institutions. By institutions I mean "organizations of people that carry out objectives using regularized practices and norms, labor, and resources" (Holland-Lulewicz et al. 2020: 1). While it is difficult to point toward exactly what these might have been and what form they took, we can say that likely they involved actions where the perceived benefits of maintaining such institutions and, by extension, the canals would have been readily apparent to the participants. Thus, the labor in the canal construction, as well as the social labor inputs to maintain not only the physical operation but also the social operation of such a built environment would facilitate a commitment to place and foster the seeking of solutions even in times of social and environmental stress (see Thompson 2016).

We can extend our thinking of collective action to systems of transport on the whole, including the caching of canoes and their production and ownership in general. Given the extensive trade systems that existed all over southern Florida, it is unlikely there were large-scale restrictions to travel within regions. Luer (1989) argues that the canal systems facilitated trade and exchange over vast areas. Thus, these too were likely more on the order of public resources for large canals and common pool resources for smaller ones that linked two settlements or a settlement with another body of water that was directly within that settlement's control or sphere of influence. In other words, smaller canals possibly had more limited users and were consistent with how we define common pool resources—that is, where use rights are limited (Blanton and Fargher 2016: 40–41; Ostrom et al. 1994: 3). But this would be difficult for larger canals except for the section nearer to and within the settlements, for example, the main section of the Pine Island canal and its subsidiaries at the Pineland site.

While difficult to substantiate archaeologically, it would seem likely that canals were either public or common pool resources that many people could use. The next logical question is, does this apply to the vessels that traversed these waterways? That is, were canoes common pool resources or private goods that people could be excluded from using? Unlike modern cars in the United States, it is doubtful that Native vessels (i.e., canoes) were a private good on the same scale. Indeed, the canoe likely lies somewhere between a private good and common pool resource depending upon the context.

In the context of the cached canoes at transit points, it is likely that these may represent more common pool resources. Obviously, they are subtractable in the sense there is not an infinite supply of canoes at these places. They would be difficult to police since no one would likely be around to stop someone who did not belong to the group from taking a canoe. Certainly, canoes could be hidden and sunk, as described in some of the Seminole histories (Carr 2012; Duggins 2019). However, this would be effective only to a point. Thus, in the case of the cached transit point canoes, they certainly could be common pool resources (see Thompson 2023a).

In contrast with cached canoes that were far away from settlements, vessels used and produced by the community or individuals probably had different rules and use rights. It is likely that such vessels were the private property of individuals or were communally owned by cooperative groups. We know that during the sixteenth century, the Calusa lived in large, multifamily houses for extended lineages and that the shift to large-scale corporate house groups may have occurred as early as AD 500 (Thompson et al. 2014). Thus, it may be that canoes at the settlement were property of such lineages and could also be passed down from one generation to the next. This makes a certain degree of sense since large-scale fishing and storage likely operated at a level much larger than the nuclear family (Thompson et al. 2020).

It is possible that Calusa leaders were able to conscript canoes for their own purposes. This is indicated by the ability to gather large groups of canoes for both defensive and offensive military action (see discussion above). Beyond this, however, there are instances where rulers order the giving of canoes to Spaniards for travel (Hann 1991: 248). And there is some indication that small groups of people or specialists engaged in vessel production, perhaps sponsored by powerful individuals or groups. Specialized production of vessels is indirectly indicated by the standardization of woodworking tools (i.e., cutting-edged shell tools) (Dietler 2008; Patton 2013: 572).

We can infer from the above discussion that the rules governing the use of waterways and canals as well as the ownership of vessels varied greatly among the Native peoples of southern Florida. For the Calusa canoes are thought to be mostly private property of individuals or corporate groups, at least for those kept and regularly used at settlements. It is possible that such rules shifted when longer-distance trade and travel were involved, when the canoes were more on the order of common pool resources. This same may have also been the case for canals; whereas long-distance canals were more open to use by different groups of people, canals that served to facilitate movement within or into a settlement may have been regulated to the settlement's inhabitants.

Discussion: Canals as the Physical and Social Armature of Political Institutions

So critical is water travel to the functioning of Native society that Bartolomé de Argüelles' account suggests that the destruction of all canoes would be the best way to disrupt Native society (Dobyns 1983: 239–240; Quinn 1979; Wheeler et al. 2003: 547–548). Water travel, vessel production, and canal construction and use were integral to the overall operation of Native American society in Florida. Given the overview and discussion of the use rights and institutions that governed watercraft, the construction of canals, and the use of waterways, what role, then, did these systems play in Calusa histories and those of other southern Florida groups?

The first point to consider is that the construction of canals occurs both broadly along the Gulf Coast and in southern Florida before any type of political centralization models commonly offered by archaeologists (i.e., chiefdoms) (Luer and Wheeler 1997; Waselkov et al. 2022). The fact that such canals occur over different parts of the region suggests that many communities had the ability to mobilize cooperative labor to create canals of considerable length. Second, even some of the larger watercraft, like the 13-m-long Weedon Island canoe (Kolianos 2019), predate to a large degree such political systems as well. Thus, the production of such vessels has a long history throughout the region.

Taken together, the long histories of canal construction and the use of vessels for long-distance trade suggests that the institutions necessary for construction, use, and governance were in place to some degree and in one form or another for hundreds and possibly more than a thousand-plus years. This is not to say that such institutions were unchanging; however, for some of these canals to be maintained over such extended periods of time required considerable long-term community investments. As I suggest above, the investment in not only the labor to construct canals but also the social ties and institutions created from them likely fostered the persistence of settlements over extended periods of time. Thus, these earlier actions and traditions of aquatic travel structured later actions and need to be considered in interpretations of political histories in the region.

What role did the canals and watercraft play in sixteenth-century Calusa political organization? It would appear the region's inhabitants constructed canals, at least the ones that have been dated before AD 1000, with at least one that postdates this time frame. Prior interpretations of canals tend to place emphases on both political and economic reasons for canals. Specifically, in addition to trade items, Luer (1989: 116) argues that canals were used for the moving of tribute to the Calusa. This indeed likely did occur, especially from the interior region of Lake Okeechobee. And it is even possible that some political leaders constructed some of the canals with this in mind; however, I argue that, unlike monumental mounds, the construction, upkeep, and use of canals was likely more inclusive than exclusive or built for the sole purpose of tribute or the glory of the elite. While this might seem a minor and obvious point, I believe it is necessary to state here as it would be easy for researchers to assume that this may be the case. Thus, most canal systems at both "the settlement level and intersettlement level represent public goods (that is the canal itself) and were used to transport public and private goods" (Thompson 2023a: 533).

While canals were likely public goods, this is not to say they played no role in the history of political centralization in southern Florida. Indeed, as Luer (1989: 114) notes, the construction and maintenance of canals would require food storage, and at "points of assemblage, storage, and redistribution of food and other items would be expected along the canals." This is an important point; if canals not only had clear benefits for people who belong to different political ranks and have varied economic interests, then the barriers to new constructions and the canals' continued upkeep would be low. Furthermore, powerful political agents could more easily use these structures for their own ends without overtly exploiting other members of a community of canal builders and users. The increasing use of canals for trade and tribute may have also influenced both canoe and canal size, as Luer points out, noting that the Pine Island canal was almost 10 m wide and was large enough to accommodate the double-hulled canoes that were described in the sixteenth and seventeenth centuries in southern Florida.

Conclusions

Rather than catalysts for or a product of political complexity, it may be better to think of canals as a result of cooperative institutions both at the intra- and intersettlement levels. Canals, along with natural waterways, served to lower transport costs in terms of the movement of physical goods as well as people and, by extension, information. The broadscale benefits conferred by these institutions served to foster long-term relationships and likely reduced conflict among different groups. While there is certainly evidence of conflict, the degree to which this is expressed along the Gulf Coast seems to be comparatively low to the inland southeast (Hutchinson 2020). These longer-term relationships, institutions, and physical infrastructure may have served as the armature of political institutions that developed along the southern Gulf Coast and ultimately created the sixteenth-century Calusa polity.

Canal infrastructure and vessel technology likely also contributed to settlement and possibly political and economic sustainability. This needs to be explored further by developing a more robust chronology of both large settlements in the region as well as identification and dating of canal systems in the region. Interestingly, in a long-term comparison of Roman road systems, Carl-Johan Dalgaard and colleagues (2022) found that greater infrastructure (i.e., road systems) was correlated in higher settlement density in 500 BC as well as greater economic activities in the 2010s. Do the Florida canal and river systems engender the same kind of benefits, or have we (EuroAmericans) modified this environment so much that the conferred good of the past has been lost to modern development? Regardless, the canal and river systems are part of the broader Indigenous histories of these Native landscapes and deserve our collective attention regarding their documentation and preservation.

References Cited

Blanton, Richard E., and Lane F. Fargher. 2016. *How Humans Cooperate: Confronting the Challenges of Collective Action.* University Press of Colorado, Boulder.

Carr, Robert S. 2012. *Digging Miami.* University Press of Florida, Gainesville.

Carr, R. S., J. Zamanillo, and J. Pepe. 2002. Archaeological Profiling and Radiocarbon Dating of the Ortona Canal (8GL4), Glades County, Florida. *Florida Anthropologist* 55:3–22.

Dalgaard, Carl-Johan, Nicolai Kaarsen, Ola Olsson, and Pablo Selaya. 2022. Roman Roads to Prosperity: Persistence and Non-Persistence of Public Infrastructure. *Journal of Comparative Economics* 50(4): 896–916.

Dietler, John Eric. 2008. Craft Specialization and the Emergence of Political Complexity in Southwest Florida. PhD dissertation, Department of Anthropology, University of California, Los Angeles.

Dobyns, Henry F. 1983. *Their Number Become Thinned: Native American Population Dynamics in Eastern North America.* University of Tennessee Press, Knoxville.

Drennan, Robert D. 1984. Long-Distance Transport Costs in Pre-Hispanic Mesoamerica. *American Anthropologist* 86(1): 105–112.

Duggins, Julie. 2019. Canoe Caching at Transit Points: Inferring Florida's Ancient Navigation Routes Using Archaeology and Ethnohistory. In *Iconography and Wetsite Archaeology of Florida's Watery Realms,* edited by Ryan J. Wheeler and Joanna Ostapkowicz, pp. 82–110. University Press of Florida, Gainesville.

Fitzpatrick, Scott M. 2013. Seafaring Capabilities in the Pre-Columbian Caribbean. *Journal of Maritime Archaeology* 8: 101–138.

Hann, John. 1991. *Missions to the Calusa.* University Press of Florida, Gainesville.

Holland-Lulewicz, Jacob, Megan Anne Conger, Jennifer Birch, Stephen A. Kowalewski, and Travis W. Jones. 2020. An Institutional Approach for Archaeology. *Journal of Anthropological Archaeology* 58: 101163.

Hutchinson, Dale L. 2020. *Bioarchaeology of the Florida Gulf Coast: Adaptation, Conflict, and Change.* University Press of Florida, Gainesville.

Knight, Vernon James, and John E. Worth. 2021. A Cuban Origin for Glades Pottery? In *Methods, Mounds, and Missions: New Contributions to Florida Archaeology,* edited by Ann S. Cordell and Jeffrey M. Mitchem, pp. 193–206. University Press of Florida, Gainesville.

Kolianos, Phyllis. 2019. Wood Preservation Dilemmas of Florida's Prehistoric Saltwater Sites. In *Iconography and Wetsite Archaeology of Florida's Watery Realms,* edited by Ryan Wheeler and Joanna Ostapkowicz, pp. 67–81. University Press of Florida, Gainesville.

Luer, George M. 1989. Calusa Canals in Southwestern Florida: Routes of Tribute and Exchange. *Florida Anthropologist* 42: 89–130.

Luer, George M., and Ryan J. Wheeler. 1997. How the Pine Island Canal Worked: Topography, Hydraulics, and Engineering. *Florida Anthropologist* 50(3): 115–131.

Marquardt, William H. 1987. The Calusa Social Formation in Protohistoric South Florida. In *Power Relations and State Formation,* edited by Thomas C. Patterson and Christine W. Gailey, pp. 98–116. Archaeology Section, American Anthropological Association, Washington, D.C.

Marquardt, William H. 1988. Politics and Production Among the Calusa of South Florida. In *Hunters and Gatherers,* Vol. 1: *History, Evolution, and Social Change,* edited by Tim Ingold, David Riches, and James Woodburn, pp. 161–188. Berg, London.

Marquardt, William H. 2013. The Pineland Site Complex: Theoretical and Cultural Contexts. In *The Archaeology of Pineland: A Coastal Southwest Florida Site Complex, AD 50–1710,* edited by William H. Marquardt and Karen J. Walker, pp. 1–22. Institute of Archaeology and Paleoenvironmental Studies, Monograph No. 4, University of Florida, Gainesville.

Marquardt, William H. 2014. Tracking the Calusa: A Retrospective. *Southeastern Archaeology* 33(1): 1–24.

Marquardt, William H., and Karen J. Walker. 2013. The Pineland Site Complex: An Environmental and Cultural History. In *The Archaeology of Pineland: A Coastal Southwest Florida Site Complex, AD 50–1710,* edited by William H. Marquardt and Karen J. Walker, pp. 793–920. Institute of Archaeology and Paleoenvironmental Studies, Monograph No. 4, University of Florida, Gainesville.

Marquardt, William H., Karen J. Walker, Victor D. Thompson, Michael Savarese, Amanda D. Roberts Thompson, and Lee A. Newsom. 2022. Episodic Complexity and the Emergence of a Coastal Kingdom: Climate, Cooperation, and Coercion in Southwest Florida. *Journal of Anthropological Archaeology* 65: 101364.

Newsom, L. A., and C. M. Scarry. 2013. Homegardens and Mangrove Swamps: Pineland Archaeobotanical Research. In *The Archaeology of Pineland: A Coastal Southwest Florida Site Complex, A.D. 50–1710,* edited by William H. Marquardt and Karen J. Walker, pp. 253–304. Institute of Archaeology and Paleoenvironmental Studies, Monograph No. 4, University of Florida, Gainesville.

Ostrom, Elinor, Roy Gardner, James Walker, James M. Walker, and Jimmy Walker. 1994. *Rules, Games, and Common-Pool Resources.* University of Michigan Press, Ann Arbor.

Patton, Rob. 2013. The Temporal Contexts of Precolumbian Shell Artifacts from Southwest Florida: A Case Study of Pineland. In *The Archaeology of Pineland: A Coastal Southwest Florida Site Complex, A.D. 50–1710,* edited by William H. Marquardt and Karen J. Walker, pp. 545–584. Institute of Archaeology and Paleoenvironmental Studies, Monograph No. 4, University of Florida, Gainesville.

Quinn, David B. 1979. *New American World,* Vol. 5: *The Extension of Settlement in Florida, Virginia, and the Spanish Southwest.* Arno Press, New York.

Rodning, Christopher B. 2003. Water Travel and Mississippian Settlement at Bottle Creek. In *Bottle Creek: A Pensacola Culture Site in South Alabama,* edited by Ian Brown, pp. 194–204. University of Alabama Press, Tuscaloosa.

Shearn, Isaac. 2020. Canoe Societies in the Caribbean: Ethnography, Archaeology, and Ecology of Precolonial Canoe Manufacturing and Voyaging. *Journal of Anthropological Archaeology* 57: 101140.

Thompson, Victor D. 2016. Finding Resilience in Ritual and History in South Florida. In *Beyond Collapse: Archaeological Perspectives on Resilience, Revitalization, and Reorganization in Complex Societies,* edited by Ronald K. Faulseit, pp. 313–341. Southern Illinois University Press, Carbondale.

Thompson, Victor D. 2023a. Considering Ideas of Collective Action, Institutions, and

"Hunter-Gatherers" in the American Southeast. *Journal of Archaeological Research* 31: 503–560.

Thompson, Victor D. 2023b. Considering Urbanism at Mound Key (Caalus), the Capital of the Calusa in the Sixteenth Century, Southwest Florida, USA. *Journal of Anthropological Archaeology* 72: 101546.

Thompson, Victor D., William H. Marquardt, Alexander Cherkinsky, Amanda D. Roberts Thompson, Karen J. Walker, Lee A. Newsom, and Michael Savarese. 2016. From Shell Midden to Midden-Mound: The Geoarchaeology of Mound Key, An Anthropogenic Island in Southwest Florida, USA. *PLOS One* 11(4): e0154611.

Thompson, Victor D., William H. Marquardt, Michael Savarese, Karen J. Walker, Lee A. Newsom, Isabelle Lulewicz, Nathan R. Lawres, Amanda D. Roberts Thompson, Allan R. Bacon, and Christoph A. Walser. 2020. Ancient Engineering of Fish Capture and Storage in Southwest Florida. *Proceedings of the National Academy of Sciences* 117(15): 8374–8381.

Thompson, Victor D., William H. Marquardt, and Karen J. Walker. 2014. A Remote Sensing Perspective on Shoreline Modification, Canal Construction and Household Trajectories at Pineland along Florida's Southwestern Gulf Coast. *Archaeological Prospection* 21(1): 59–73.

Thompson, Victor D., William H. Marquardt, Karen J. Walker, Amanda D. Roberts Thompson, and Lee A. Newsom. 2018. Collective Action, State Building, and the Rise of the Calusa, Southwest Florida, USA. *Journal of Anthropological Archaeology* 51: 28–44.

Thompson, Victor D., and Thomas J. Pluckhahn. 2012. Monumentalization and Ritual Landscapes at Fort Center in the Lake Okeechobee Basin of South Florida. *Journal of Anthropological Archaeology* 31(1): 49–65.

Thompson, Victor D., Amanda D. Roberts Thompson, and John E. Worth. 2018. The Political Ecology of the Event in Calusa Colonial Entanglements. In *Uneven Terrain: Archaeologies of Political Ecology,* edited by John K. Millhauser, Christopher T. Morehart, and Santiago Juarez, pp. 68–82. Archaeological Papers of the American Anthropological Association, Washington, D.C.

Thompson, Victor D., and John E. Worth. 2011. Dwellers by the Sea: Native American Adaptations along the Southern Coasts of Eastern North America. *Journal of Archaeological Research* 19(1): 51–101.

Waselkov, Gregory A., Donald A. Beebe, Howard Cyr, Elizabeth L. Chamberlain, Jayur Madhusudan Mehta, and Erin S. Nelson. 2022. History and Hydrology: Engineering Canoe Canals in the Estuaries of the Gulf of Mexico. *Journal of Field Archaeology* 47(7): 486–500.

Wheeler, Ryan J. 1995. The Ortona Canals: Aboriginal Canal Hydraulics and Engineering. *Florida Anthropologist* 48: 265–281.

Wheeler, Ryan J. 1998. Walker's Canal: An Aboriginal Canal in the Florida Panhandle. *Southeastern Archaeology* 17(2): 174–181.

Wheeler, Ryan J., James J. Miller, Ray M. McGee, Donna Ruhl, Brenda Swann, and Melissa Memory. 2003. Archaic Period Canoes from Newnans Lake, Florida. *American Antiquity* 68(3): 533–551.

Widmer, Randolph J. 1988. *The Evolution of the Calusa: A Non-Agricultural Chiefdom on the Southwest Coast of Florida.* University of Alabama Press, Tuscaloosa.

Worth, John. 2013. Pineland during the Spanish Period. In *The Archaeology of Pineland: A Coastal Southwest Florida Site Complex, A.D. 50–1710,* edited by William H. Marquardt and Karen J. Walker, pp. 767–792. Institute of Archaeology and Paleoenvironmental Studies, Monograph No. 4, University of Florida, Gainesville.

Worth, John. 2014. *Discovering Florida: First-Contact Narratives of Spanish Expeditions along the Lower Gulf Coast.* University Press of Florida, Gainesville.

8

The Transformative Power of Boats

Seafaring, Maritime Interaction, and Social Complexity

MIKAEL FAUVELLE AND PETER JORDAN

The use of technology is one of the key characteristics of our species. Techno-logical innovations have allowed humans to spread to all corners of the globe and have provided us with the tools to survive and thrive in a diverse variety of landscapes (Frieman 2021; Jordan 2015). One of the best examples of the importance of technology in shaping human capacities can be found in mari-time and coastal societies, where technological innovations have allowed us to exploit aquatic environments vastly different from those in which we evolved. Fishhooks, nets, harpoons, sails, and many other technological innovations are the tools that allowed our ancestors to build rich and resilient maritime societies in coastal and island areas around the world. Of all of these innova-tions, perhaps none has been as important as the boat itself, which literally allowed ancient people to navigate the watery worlds that surrounded them. The innovation of these boats, we suggest, had profound implications for the societies that used them, shaping new social possibilities, and altering the path of historical trajectories.

In this chapter, we explore the relationship between watercraft innovation and the development of expansive systems of social inequality. We argue that the development of certain types of boats—namely, complex composite sewn canoes—created positive feedback systems that encouraged additional invest-ment in maritime networks and lifeways while funneling resources into the hands of boat builders and boat owners. We work from the observation that in many different societies around the world (but not all: see chapters in this vol-ume by Aguilera et al. [6] and García-Piquer [2]), the innovation of sewn-plank canoes was temporally correlated with the expansion of Indigenous maritime

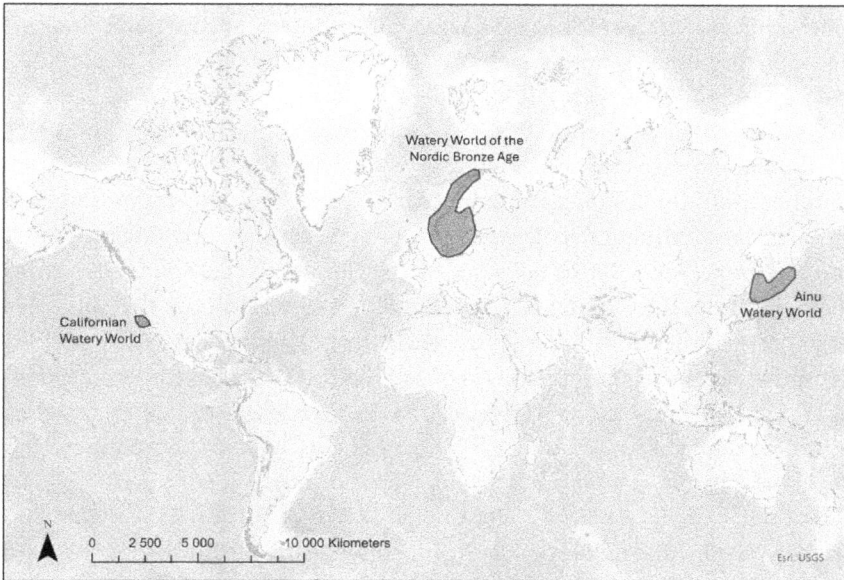

Figure 8.1. Map showing regions discussed in text.

networks and the development of systems of political hierarchy (Fauvelle et al. 2024). After laying the theoretical basis for our hypothesized connection between watercraft innovation and social complexity, we explore its implications by comparing case studies from California, Scandinavia, and East Asia (Figure 8.1).

Watercraft, Technology, and Social Complexity

Technological traditions are embedded social institutions that provide solutions to problems encountered by human societies while also shaping and transforming social capacities and historical trajectories (Jordan 2015). The innovation of new technologies, including material traditions such as ceramics or stone tools or social technologies such as writing or money, is a key mechanism through which social change occurs (Fauvelle 2024; Frieman 2021; Mann 1986; Pfaffenberger 1992). Watercraft are central to the technological traditions of maritime societies as they provide new material capacities while also shaping social traditions such as through the need to crew and captain boats (Ames 2002; Arnold 1995; Fauvelle 2011; Ling, Earle et al. 2018). In a previous article,

Fauvelle, Sasaki, and Jordan (2024) explored the relationship between watercraft innovation and the formation of expansive Indigenous world systems. Here we refine this argument to focus on the development of systems of social hierarchy and expand our comparison to incorporate ancient Scandinavia.

As many other chapters in this volume emphasize, much of the groundwork for the themes explored in this chapter was laid out by the foundational research of Ken Ames. In his seminal paper *Going by Boat,* Ames described the logistical implications of boat ownership for what he termed as aquatic hunters and gatherers (Ames 2002). Focusing on the Northwest coast but also drawing comparisons to the Jomon and the Chumash, Ames points out that the use of boats creates different realities for the settlement mobility of maritime peoples. The ability to move large amounts of goods reduces the needs for remote logistics camps and leads to larger, more centralized, and more sedentary societies. Without a doubt, Ames showed that the use of boats matters for anthropological theory of maritime peoples (Ames 2002: 20).

But not all boats are the same. Ames pointed this out as well, showing how umiaks, kayaks, and dugouts all had different logistic capabilities (Ames 2002: 26–30). Different types of boats, however, also had different social implications for the societies that used them. While some types of boats—such as simple dugouts or reed craft—were relatively easy to construct, others involved entire communities and demanded the assemblage of considerable material resources. Often these larger, faster, and more capable boats became important social symbols and markers of status for the societies that used them (Chacon et al. 2020; Fauvelle 2011; Fauvelle et al. 2024). It is these more complex watercraft that are the focus of this chapter.

We suggest that "complex watercraft" are those that required resources beyond the capacity of a household or small group to assemble. These resources can include specialized or esoteric knowledge, imported or valuable materials, or high investments of labor. Following this definition, complex watercraft can take lots of different forms—they can be massive, steam-formed dugouts like the Haida canoes of the Northwest coast (Arima 1983; Ling, Chacon et al. 2018; Moss 2008) or composite boats built from resins, reeds, and log cores (Des Lauriers 2005; Fauvelle et al. 2024; Ohtsuka 1999). In this chapter we focus on a specific type of complex craft—the sewn-plank canoe—and explore how innovation in boat designs led to runaway increases in the social complexity of the societies that built them.

Our proposed model works as follows: Complex watercraft were originally innovated to solve logistical problems unrelated to issues of social inequality—this can include transportation distances, speed, the need to cross dangerous

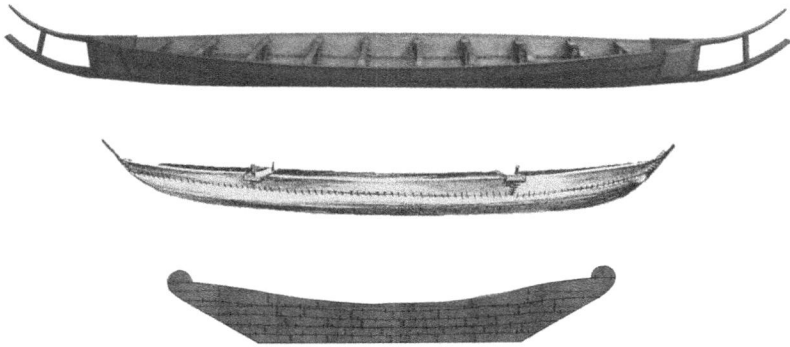

Figure 8.2. Comparison of three different sewn-boat technologies. *Top*: Model of the Hjortspring boat from Denmark, prepared by Richard Potter. *Middle*: Late nineteenth-century diagram of a traditional Ainu sewn-plank boat. Public domain image. Source: Landor 2012/1893. *Bottom*: Depiction of Chumash plank canoe. Drawing by Mikael Fauvelle.

seas, cargo capacity needs, or a combination of all of the above. Once the use of these new watercraft took hold, however, the higher resource cost needed to build them led to their monopolization by those with the access to those resources. This monopoly included not only the construction of boats but rewards that came with their use for long-distance trade and travel. Quickly, this led to a positive feedback loop where those who financed boat construction accumulated resources at a much greater rate than their non-boat-owning peers. Soon boat-owning families or lineages were the only ones that could afford to build more boats and reap the benefits that boat-owning entails. Where small social differences may have originally existed, social hierarchies soon formed. Complex boats therefore provided a mechanism by which social inequality developed as an unintended consequence of initially attractive and advantageous technological innovations.

As with all models, we do not mean to suggest that this was the pathway to complexity followed by maritime societies. We do believe, however, that it matches the historical trajectories taken by many early complex coastal societies around the world. In the rest of this chapter, we explore three different case studies where we argue that the innovation of a specific kind of complex boat—namely, sewn-plank canoes—eventually led to wide-ranging social changes.

California

Indigenous societies in precolonial southern California provide an excellent case study in the connection between watercraft use and the origins of political and economic complexity. During the time of Spanish contact, California's Channel Islands and the mainland adjacent coast were the homeland of the Chumash and Tongva people, who were maritime fisher-foragers living in sedentary villages characterized by political hierarchies and economic inequality (Arnold 2001; Fauvelle and Somerville 2024; Gamble 2008; Kennett 2005). The Chumash and Tongva used shell beads for money, which underwrote an intense and active trade economy with connections across the American Southwest (Fauvelle 2024, 2025; Fauvelle and Perry 2019, 2023; Gamble 2020; Smith and Fauvelle 2015). To move between islands and from the islands to the mainland, the Chumash relied on seaworthy boats to ferry both goods and people.

The California plank canoe was built out of redwood or pine logs lashed together with milkweed fiber and caulked with a mixture of pine pitch and the natural tar that seeps from the ground in many locations across southern California (Hudson et al. 1978; see Figure 8.2). These boats were built by both the Chumash and Tongva people and were called *tomol* in Chumash and *ti'at* in Tongva. Much of what we know about the plank canoe comes from one built in 1912 for anthropologist J. P. Harrington by Fernando Librado (Kit-se-pa-wit), who designed the canoe based on his recollections of boats built when he was a young boy in the 1840s and 1850s (Hudson et al. 1978). While other boats such as dugouts and reed boats were used by the Chumash and Tongva, the speed and cargo capacity of the plank canoe made it by far the most practical and preferred vessel for interisland voyages and trade (Fauvelle and Montenegro 2024).

The importance of the plank canoe for the political economy of the Chumash region cannot be understated. As has been highlighted by Jeanne Arnold (1995), ownership over plank canoes would have been central to the activities of entrepreneurial chiefs who sought to corner the exchange of shell beads and other goods across the Santa Barbara Channel (Fauvelle and Ling 2025). Material evidence in the form of drills, canoe plugs, and redwood planks date the initial use of the plank canoe to around 500 CE (Gamble 2002), a time of increasing social inequality in the channel region that cumulated with the formation of incipient chiefdoms several centuries later (Arnold 2001). The impact of the innovation of the plank canoe can even be seen in the regions' genetic history, which shows an increase in available mate pools and subsequent gene flow on the islands after around 500 CE (Nakatsuka et al. 2023: 128).

Unlike previously used dugouts or reed boats, plank canoes required several high-status and difficult to acquire materials (Fauvelle 2011, 2014; Hudson et al. 1978). The boat planks themselves were often made from redwood, which is not native to southern California and was collected from driftwood on the region's shores. Pine was sometimes substituted but is much more abundant on the mainland than on the islands. Milkweed fiber, needed to lash together the boat's planks, is rare on the Channel Islands, as is high-quality tar for caulking. In fact, Librado specifically stated that only tar sourced from a select few mainland mains was of high enough quality to build a *tomol* (Fauvelle 2011; Hudson et al. 1978). It should not be surprising that many of these locations were also home to some of the largest and most powerful Chumash chiefly centers.

The knowledge needed to build a plank canoe was also carefully controlled. Canoe builders were members of the Brotherhood of the Tomol, which restricted membership to individuals from high-status family lineages (Hudson et al. 1978). The labor involved in boat construction was enormous as well. It is estimated that up to 540 hours of labor went into the construction of a single *tomol* spanning up to six months of work. Considering the high value of the materials that went into their construction, the esoteric nature of the knowledge needed to build them, and the massive amounts of labor required, it is unsurprising that only high-status and elite individuals were able to construct and own plank canoes. Although boat owners sometimes allowed others to use their canoes, they expected a portion of any profits in return, further concentrating wealth into their hands (Hudson et al. 1978: 130).

The intensive exchange and social transformations allowed for by the innovation of the plank canoe led to widespread impacts in southern California and beyond. As plank canoes could only be built by elites, these individuals quickly came to dominate trade across the Santa Barbara Channel. And trade was enormous. In the centuries between around 800 CE and the time of Spanish conquest, many millions of shell beads were produced on the Channel Islands off the coast of California and traded to the mainland coast and beyond across the interior regions of the American West (Fauvelle 2024, 2025; Gamble 2020; Smith and Fauvelle 2015). This intensive trade would not have been possible without the development of the plank canoe. Considering that Chumash shell beads were used as currency as far as modern New Mexico (Bennyhoff and Hughes 1987), innovation in watercraft technology in coastal California therefore had continent-spanning implications of socioeconomic systems in ancient North America.

Scandinavia

The abundant islands and waterways of Scandinavia have made the region a center for maritime traditions throughout the history of its human occupation. Mesolithic and Neolithic Scandinavian seafarers used a wide range of boat types including dugouts, bark boats, and possibly skin boats to navigate the waters of the Baltic and North Seas (Jansson 2004; Lanting 1997; von Arbin and Lindberg 2017). During the Bronze Age, however, Scandinavian seafarers began to experiment with a radical new type of boat—the sewn canoe. Built from drilled planks lashed together with cordage and rawhide and caulked with pitch and tallow, the sewn canoe represented a radical change from previous boat-building technologies (Crumlin-Pedersen and Trakadas 2003). Up to 20 m long and 2 m wide, Bronze Age sewn canoes could rapidly transport dozens of people and large hauls of cargo over large distances, allowing for a form of seafaring society that would shape Scandinavia's future for millennia to come.

The best example of a Scandinavian sewn canoe is the spectacular Hjortspring boat, excavated in the 1920s and now displayed at the National History Museum in Copenhagen (Crumlin-Pedersen and Trakadas 2003; Figure 8.2). The Hjortspring boat dates to the early Iron Age—around the third century BCE—but strong similarities between the boat and Bronze Age rock art suggest that it represents a much older boat-building tradition. The oldest image of a boat of the Hjortspring type dates to just after the beginning of the Scandinavian Bronze Age at 1600 BCE and is found carved in bronze on one of the Rørbyswords (Crumlin-Pedersen 2003). Other examples of sewn-boat fragments have been excavated at sites across Scandinavia but are much more fragmentary than the intact Hjortspring find. For example, the Hampnäs thwart, discovered in northern Sweden in the 1920s and later carbon dated to the early fourth century BCE, is almost identical to similar thwarts on the Hjortspring boat (Ramqvist 2010).

It is clear that sewn canoes were used from at least the start of the Bronze Age, and their development and use likely went hand in hand with the sweeping social changes that came to define this period (Horn et al. 2024; Ling, Chacon et al. 2018; Ling, Earle et al. 2018). The defining technology of the age—bronze—required tin imported from the British Isles or Iberia (Ling et al. 2013, 2014). Voyages to these distant destinations would have been exceedingly difficult in a dugout but would have been achievable in a Hjortspring-like boat. Even a round trip from Jutland to Iberia and back again would have only taken around three months at average speeds in a plank boat, possibly allowing such trips to be conducted during periods of good weather. The use of plank canoes, therefore, can be seen as a precursor for the changes that came to define the Nordic Bronze Age.

The social implications of plank canoes would also have been huge. Plank canoes would have required considerably greater skills, resources, and labor investments than previous boats. This likely meant that their use would have been sponsored by elite individuals. These leaders can even be seen in rock art depictions of Bronze Age boats, standing taller and larger than their rowing compatriots (Chacon et al. 2020; Horn et al. 2024). Sewn boats would also have allowed for the transport of much larger groups of people, moving small armies that could be used for raiding and warfare—and, critically, for transporting slaves and captured goods back home. These benefits would quickly have accrued into substantial wealth, as can be seen in the numerous gold hordes that define the period. This "maritime mode of production," which started with the innovation of plank boats, continued to define Scandinavia through the end of the Viking Age (Fauvelle and Ling 2025; Ling, Earle et al. 2018).

The case studies from California and Scandinavia provide clear parallels in the importance of watercraft innovation for the formation of political hierarchies. In both areas, the innovation of sewn-plank-boat technology allowed for the expansion of maritime trade networks that eventually came to define the political economy of the region. Entrepreneurial chiefs in both Scandinavia and coastal California sponsored voyages that brought them great wealth, which they then were able to funnel into the construction of more boats and the financing of additional voyages. Some differences do exist between these case studies. Raiding for slaves was a central component of maritime activity in the Nordic Bronze Age and was absent in precolonial California. Scandinavian sewn boats were also substantially bigger than those used in California, despite close similarities in the sewn-plank technology used to construct them. Nonetheless, without the use of these advanced watercraft, it is unlikely that the network-oriented chiefly systems that came to define the Late Period Chumash and the Nordic Bronze Age would have come about.

East Asia

Our final example takes us back to the Pacific to the island of Hokkaido and the Sea of Okhotsk. Here Indigenous Ainu people also built a form of sewn-plank canoe, the *itaomachip* (Fitzhugh and Luukkanen 2019; see also Figure 8.2). These plank boats were built from a dugout core made from massive katsura (*Cercidiphyllum* sp.) logs, with raised sides made from sewing together pine boards with hemp rope (Ohtsuka 1999). Like the other sewn boats discussed in this chapter, these plank boats were also caulked with pine pitch. The *itaomachip* was slightly larger than the *tomol* or *ti'at* and smaller than the Hjortspring-style Nordic plank boat. Unlike these other examples, the *itaomachip* was built from a dugout hull, but the sewn-plank technology was otherwise quite similar. Another difference is that the *itaomachip* was often equipped with a sail, technology that was not used in California and may have been absent in Bronze Age Scandinavia (cf. Bengtsson 2015). All three examples, however, were specialized seafaring boats that took their occupants on long-distance trading voyages in the open ocean.

Using the *itaomachip*, Ainu seafarers were able to trade and raid across a vast area of maritime Northeast Asia, including southward into much of Honshu, but also northward in the Kurils, and from Sakhaklin island into the major arterial waterways of the Amur River basin (Fauvelle et al. 2024; Hudson 2014; Yamaura 1999). Use of the *itaomachip* enabled the Ainu to emerge as a semiautonomous maritime culture that could operate across vast intercultural worlds of a scale and diversity that was broadly analogous to those initiated by sewn canoe use in Bronze Age Scandinavia and the Chumash interaction sphere of western North America (Hudson 2013, 1999: 206–232; Koji 2013; Walker 2001). The political economy of all three regions is also structurally similar, as the proceeds from raiding and opportunistic trading would have been funneled back into the more localized, status-building agendas and practices of aspiring Ainu leaders. Here the acquisition and ritualized display and consumption of exotic trade goods was central to manipulation of social networks and the concentration of wealth into particular families and lineages (Walker 2001), a process also seen among emergent Chumash and Scandinavian chiefs. In fact, the deeper historical emergence of these dynamics among the Ainu almost certainly extended further back in time and into the complex constellation of antecedent archaeological cultures including the Jomon, Epi-Jomon, Satsumon, and especially Okhotsk traditions (Hudson 1999: 206–232; Walker 2001). This emerging pattern can be envisaged as a maritime mode of production that is strikingly similar to that seen in Bronze Age Scandinavia (Hudson 2022).

For particular reasons, the historiography of Ainu ethnogenesis, their lifeways, and their traditions has overwhelmingly sought to understand these communities as "relict" hunting-fishing-gathering peoples who had somehow survived into the ethnographic period (Hudson 2013, 2014, 1999: 207; Koji 2013; Lewallen 2017). As a result, the historical role of the Ainu as an autonomous maritime-oriented trading and voyaging culture has only recently started to emerge, and many key aspects of these lifeways, including innovation of oceangoing boat technologies need further research (Fauvelle et al. 2024). These interpretive biases and lacunae are further compounded by limited physical and artistic depictions of early boats, making it difficult to pin down precisely when the larger sewn-plank vessels started to be used in Northeast Asia. One strong possibility, however, is that Ainu canoe technologies emerged out of vessels used by the preceding Okhotsk culture (Yamaura 1998), which appeared in Hokkaido as part of a wider series of population movements and maritime expansions linked to new patterns of trade and interaction involving both continental and agrarian states and empires, on the one hand, and new kinds of maritime networks and peripheries, on the other, together making up what might be defined as an "East Asian World System" (Hudson 1999: 207–232; Yamaura 1998, 1999).

Among the Okhotsk culture, expansive patterns of interregional and intercultural trade and exchange relations add interpretive challenges and classificatory nuances to defining and understanding sedentary communities that occupied large coastal villages and subsisted primarily on fishing, marine mammal hunting, and small-scale pig-rearing. They form classic coastal foragers, but their sedentism and social complexity (inequality) has analogs with the Pacific Northwest Coast (Prentiss 2023) as well as with the Chumash example from California. The extent to which the Okhotsk culture versus the more inland-facing and horticulture-, fishing-, and hunting-oriented Satsumon culture contributed to the emergence of the Ainu after around AD 1200 is a long debate (Hudson 1999; Walker 2001), but many of the Ainu maritime practices—perhaps including boat technologies—and some aspects of worldview like the bear festival may indeed be traced back to the former, although there is emerging consensus that the Satsumon were a more direct ancestor to the Ainu.

If the Ainu use of the *itaomachip* did extend back to the Okhotsk culture, then it must have developed within in dynamic interaction and interregional exchange and trading networks that have striking parallels with the other two case studies presented in this chapter. More work will be needed to better understand the exact developmental process of the sewn canoe in Hokkaido and how it relates to Ainu ethnogenesis, viewed not as relict inland hunters and fishers of Japan's northernmost island "wilderness" but as an expansive and independent and autonomous culture that acted and interacted with a wide array of surrounding cultures and polities. It is clear that voyaging affordances of the *itaomachip* were a powerful driver that enabled the historical Ainu to operate in analogous ways to other plank canoes found in complex Indigenous maritime societies in other world regions. In all three case studies discussed in this chapter, therefore, we can see a close and dialectic relationship among individual aspiration, status-building agendas, cumulative boat innovation, and the wider political economic systems focused on network-oriented trading and the accumulation of wealth and resources.

Conclusions

We would like to return to some comparisons with some of the foundational concepts developed by Ken Ames, which were discussed at the start of this chapter. For Ames (2002), the societal and human behavioral importance of boats lay in their sheer logistical capacity to shift people, equipment, and resources on scales undreamed of by pedestrian foragers. The power of boats transforms many of the basic options and abilities open to terrestrial hunter-gatherers as they seek to move people to resources, but this power also shifts resources back to people settled in places where the resources can be cached, shared, and consumed. We have focused on a different but—we argue—equally important aspect of boats: the ability of boat owners to use their boats to manipulate and monopolize economic activity. By financing boats, elites would reap enormous rewards that would not necessarily have been apparent to those who initially experimented with boat innovation. The wealth accrued through maritime activity that could then be consolidated into households and lineages over generations, leading to the formation of an entrenched elite. Experiments with boat technology would thus have kicked off a domino chain of changes that had the potential to entrench social inequality while also affecting widespread interaction networks due to increased contact and exchange with maritime peoples.

We hope that these comparisons have also reiterated another one of Ames' points—that boats matter for anthropological theory. They matter not only for theory regarding hunter-gatherer logistics but also for our theories of social evolution and change in early complex societies around the world. While most progressive linear narratives of the origins of social complexity focus on terrestrial and agrarian routes to surplus, inequality, and political hierarchy, the three case studies discussed here clearly followed a distinctly different path that transcends forager-farmer-states. We hope that future research on routes in the use of watercraft in maritime societies will continue to fill in details on the mechanisms driving the formation of systems of political and economic complexity in ancient societies around the world.

References Cited

Ames, K. M. 2002. Going by Boat: The Forager-Collector Continuum at Sea. In *Beyond Foraging and Collecting: Evolutionary Change in Hunter-Gatherer Settlement Systems*, edited by Ben Fitzhugh and Junko Habu, pp. 19–52. Kluwer Academic/Plenum Publishers, New York.

Arima, Eugene Y. 1983. *The West Coast People: The Nootka of Vancouver Island and Cape Flattery*. British Columbia Provincial Museum, Victoria.

Arnold, Jeanne E. 1995. Transportation Innovation and Social Complexity among Maritime Hunter Gatherer Societies. *American Anthropologist* 97(4): 733–747.

Arnold, Jeanne E. 2001. *The Origins of a Pacific Coast Chiefdom: The Chumash of the Channel Islands.* University of Utah Press, Salt Lake City.

Bengtsson, Boel. 2015. Sailing Rock Art Boats: A Reassessment of Seafaring Abilities in Bronze Age Scandinavia and the Introduction of the Sail in the North. PhD dissertation, Department of Archaeology, University of Southampton, Southampton.

Bennyhoff, J. A., and R. E. Hughes. 1987. *Shell Bead and Ornament Exchange Networks between California and the Western Great Basin.* Anthropological Papers Vol. 64. American Museum of Natural History, New York.

Chacon, Richard, Johan Ling, Brian Hayden, and Yamilette Chacon. 2020. Understanding Bronze Age Scandinavian Rock Art: The Value of Interdisciplinary Approaches. *Adoranten* (December): 74–95.

Crumlin-Pedersen, Ole. 2003. The Hjortspring Boat in a Ship-Archaeological Context. In *Hjortspring: A Pre-Roman Iron-Age Warship in Context,* edited by Ole Crumlin-Pedersen and Athena Trakadas, pp. 209–232. Viking Ship Museum, Roskilde.

Crumlin-Pedersen, Ole, and Athena Trakadas (editors). 2003. *Hjortspring: A Pre-Roman Iron-Age Warship in Context.* Viking Ship Museum, Roskilde.

Des Lauriers, Matthew R. 2005. The Watercraft of Isla Cedros, Baja California: Variability and Capabilities of Indigenous Seafaring Technology along the Pacific Coast of North America. *American Antiquity* 70(2): 342–360.

Fauvelle, Mikael. 2011. Mobile Mounds: Asymmetrical Exchange and the Role of the Tomol in the Development of Chumash Complexity. *California Archaeology* 3(2): 141–158.

Fauvelle, Mikael. 2014. Acorns, Asphaltum, and Asymmetrical Exchange: Invisible Exports and the Political Economy of the Santa Barbara Channel. *American Antiquity* 79(3): 573–575.

Fauvelle, Mikael. 2024. *Shell Money: A Comparative Study.* Cambridge University Press, Cambridge.

Fauvelle, Mikael. 2025. The Trade Theory of Money: External Exchange and the Origins of Money. *Journal of Archaeological Method and Theory* 32, art. 23. https://doi.org/10.1007/s10816-025-09694-9.

Fauvelle, Mikael, and Johan Ling. 2025. Larger Boats, Longer Voyages, and Powerful Leaders: Comparing Maritime Modes of Production in Scandinavia and California. In *Maritime Encounters I: Presenting Counterpoints to the Dominant Terrestrial Narrative of European Prehistory,* edited by John T. Koch, Mikael Fauvelle, Barry Cunliffe, and Johan Ling. Oxbow Books, Oxford.

Fauvelle, Mikael, and Alvaro Montenegro. 2024. Do Stormy Seas Lead to Better Boats? Exploring the Origins of the Southern Californian Plank Canoe Through Ocean Voyage Modeling. *Journal of Island & Coastal Archaeology* 20(3): 1–21. https://doi.org/10.1080/15564894.2024.2311107.

Fauvelle, Mikael, and Jennifer Perry. 2019. Material Conveyance and Trade in the Channel Region. In *An Archaeology of Abundance: Reevaluating the Marginality of California's Islands,* edited by Kristina M. Gill, Mikael Fauvelle, and Jon M. Erlandson, pp. 191–225. University Press of Florida, Gainesville.

Fauvelle, Mikael, and Jennifer Perry. 2023. Fisher-Hunter-Gatherer Complexity on Cali-
fornia's Channel Islands: Feasting, Ceremonialism, and the Ritual Economy. In *Fisher-Hunter-Gatherer Complexity in North America,* edited by Christina Perry Sampson,
pp. 194–224. University Press of Florida, Gainesville.

Fauvelle, Mikael, Shiro Sasaki, and Peter Jordan. 2024. Maritime Technologies and
Coastal Identities: Seafaring and Social Complexity in Indigenous California and
Hokkaido. *Indigenous Studies and Cultural Diversity* 1(2): 30–52.

Fauvelle, Mikael, and Andrew Somerville. 2024. Diet, Status, and Incipient Social Inequality: Stable Isotope Data from Three Complex Fisher-Hunter-Gatherer Sites in
Southern California. *Journal of Anthropological Archaeology* 73: 101554. https://doi
.org/10.1016/j.jaa.2023.101554.

Fitzhugh, William W., and Harri T. Luukkanen. 2019. The Indigenous Watercraft of
Northern Eurasia. *Vestnik of Saint Petersburg University: History* 64(2): 474–489.
https://doi.org/10.21638/11701/spbu02.2019.206.

Frieman, Catherine J. 2021. *An Archaeology of Innovation: Approaching Social and Techno-logical Change in Human Society.* Manchester University Press, Manchester.

Gamble, Lynn H. 2002. Archaeological Evidence for the Origin of the Plank Canoe in
North America. *American Antiquity* 67(2): 301–315.

Gamble, Lynn H. 2008. *The Chumash World at European Contact: Power, Trade, and
Feasting among Complex Hunter-Gatherers.* University of California Press, Berkeley.

Gamble, Lynn H. 2020. The Origin and Use of Shell Bead Money in California. *Journal of
Anthropological Archaeology* 60: 101237.

Horn, Christian, Knut Ivar Austvoll, Magnus Artursson, and Johan Ling. 2024. *Nordic
Bronze Age Economies.* Elements in Ancient and Premodern Economies. Cambridge
University Press, Cambridge.

Hudson, Mark. 2022. *Bronze Age Maritime and Warrior Dynamics in Island East Asia.*
Cambridge University Press, Cambridge.

Hudson, Mark J. 1999. *Ruins of Identity: Ethnogenesis in the Japanese Islands.* University of
Hawaii Press, Honolulu.

Hudson, Mark J. 2013. Ainu and Hunter-Gatherer Studies. *Beyond Ainu Studies: Changing Academic and Public Perspectives,* edited by Mark J. Hudson, Ann-Elise Lewallen,
and Mark K. Watson, pp. 117–135. University of Hawai'i Press, Hawai'i Scholarship
Online, Honolulu.

Hudson, Mark J. 2014. The Ethnohistory and Anthropology of "Modern" Hunter-
Gatherers: North Japan (Ainu). In *The Oxford Handbook of the Archaeology and
Anthropology of Hunter-Gatherers,* edited by Vicki Cummings, Peter Jordan, and
Marek Zvelebil, pp. 1054–1070. Oxford Academic, Online ed. https://doi.org/10.1093/
oxfordhb/9780199551224.001.0001.

Hudson, Travis, Janice Timbrook, and Melissa Rempe. 1978. *Tomol: Chumash Watercraft
as Described in the Ethnographic Notes of John P. Harrington.* Ballena, Socorro.

Jansson, Seth. 2004. Norrländska fynd med belysning på hällristningstid. *Botnisk Kontakt*
11: 101–112.

Jordan, Peter David. 2015. *Technology as Human Social Tradition.* University of California
Press, Berkeley.

Kennett, Douglas J. 2005. *The Island Chumash: Behavioral Ecology of a Maritime Society.* University of California Press, Berkeley.

Koji, Deriha. 2013. Trade and the Paradigm Shift in Research on Ainu Hunting Practices. In *Beyond Ainu Studies: Changing Academic and Public Perspectives,* edited by Mark J. Hudson, Ann-Elise Lewallen, and Mark K. Watson, pp. 136–150. University of Hawai'i Press, Hawai'i Scholarship Online, Honolulu.

Lanting, Jan N. 1997. Dates for Origin and Diffusion of the European Logboat. *Palaeohistoria* 39/40: 627–650.

Lewallen, Ann-Elise. 2017. The Fabric of Indigeneity: Ainu Identity, Gender, and Settler Colonialism in Japan. *Fourth World Journal* 15(2): 95–97.

Ling, Johan, Richard Chacon, and Yamilette Chacon. 2018. Rock Art, Secret Societies, Long-Distance Exchange, and Warfare in Bronze Age Scandinavia. In *Prehistoric Warfare and Violence: Quantitative and Qualitative Approaches,* edited by Andrea Dolfini, Rachel J. Crellin, Christian Horn, and Marion Uckelmann, pp. 149–174. Springer, Cham.

Ling, Johan, Timothy Earle, and Kristian Kristiansen. 2018. Maritime Mode of Production: Raiding and Trading in Seafaring Chiefdoms. *Current Anthropology* 59(5): 469–662.

Ling, Johan, Eva Hjärthner-Holdar, Lena Grandin, Kjell Billström, and Per-Olof Persson. 2013. Moving Metals or Indigenous Mining? Provenancing Scandinavian Bronze Age Artefacts by Lead Isotopes and Trace Elements. *Journal of Archaeological Science* 40(1): 291–304.

Ling, Johan, Zofia Stos-Gale, Lena Grandin, Kjell Billström, Eva Hjärthner-Holdar, and Per-Olof Persson. 2014. Moving Metals II: Provenancing Scandinavian Bronze Age Artefacts by Lead Isotope and Elemental Analyses. *Journal of Archaeological Science* 41 (January): 106–132.

Mann, Michael. 1986. *The Sources of Social Power. Vol. 1: A History of Power from the Beginning to AD 1760.* Cambridge University Press, Cambridge.

Moss, Madonna L. 2008. Islands Coming Out of Concealment: Traveling to Haida Gwaii on the Northwest Coast of North America. *Journal of Island and Coastal Archaeology* 3(1): 35–53.

Nakatsuka, Nathan, Brian Holguin, Jakob Sedig, Paul E. Langenwalter, John Carpenter, Brendan J. Culleton, Cristina García-Moreno, et al. 2023. Genetic Continuity and Change among the Indigenous Peoples of California. *Nature* 624(7990): 122–129.

Ohtsuka, K. 1999. Itaomachip: Reviving a Boat Building and Trading Tradition. In *Ainu: Spirit of a Northern People,* edited by Letitia O'Connor and Sherri Schottlaender, pp. 374–376. Arctic Studies Center, National Museum of Natural History, Smithsonian Institution, in association with University of Washington Press [Seattle], Washington D.C.

Pfaffenberger, Bryan. 1992. Social Anthropology of Technology. *Annual Review of Anthropology* 21: 491–516.

Prentiss, Anna Marie. 2023. *Ancient and Pre-Modern Economies of the North American Pacific Northwest.* Elements in Ancient and Pre-Modern Economies. Cambridge University Press, Cambridge.

Ramqvist, Per. 2010. *Lokaliseringen av fyndplatsen för den s k Hampnästoften våren 2001*. UMARK. Institutionen för idé-och samhällsstudier, Umeå universitet.

Smith, Erin, and M. Fauvelle. 2015. Regional Interactions between California and the Southwest: The Western Edge of the North American Continental System. *American Anthropologist* 17(4): 710–721.

von Arbin, Staffan, and Maria Lindberg. 2017. News on the Byslätt Bark "Canoe." In *Baltic and Beyond: Change and Continuity in Shipbuilding. Proceedings of the Fourteenth International Symposium on Boat and Ship Archaeology,* edited by J. Litwin, pp. 245–250. National Maritime Museum, Gdansk.

Walker, Brett L. 2001. *The Conquest of Ainu Lands: Ecology and Culture in Japanese Expansion, 1590–1800*. University of California Press, Berkeley.

Yamaura, Kiyoshi. 1998. The Sea Mammal Hunting Cultures of the Okhotsk Sea with Special Reference to Hokkaido Prehistory. *Arctic Anthropology* 35(1): 321–334.

Yamaura, Kiyoshi. 1999. Prehistoric Hokkaido and Ainu Origins. In *Ainu: Spirit of a Northern People,* edited by Letitia O'Connor and Sherri Schottlaender, pp. 39–46. Arctic Studies Center, National Museum of Natural History, Smithsonian Institution, in association with University of Washington Press [Seattle], Washington, D.C.

9

Going by Boat-Being

An Indigenous Ontological Approach to Watercraft in the Pacific Northwest Coast

ERIN M. SMITH

The title of this chapter is inspired by Ken Ames and his seminal paper on watercraft, titled "Going by Boat" (2002). In the paper, Ames addressed the theoretical importance of boats among aquatic hunter-gatherers and showed how the integration of boats was essential to their daily lives. *Going by Boat* was truly an important contribution not only to Pacific Northwest Coast (PNWC) archaeology but also to scholarship on aquatic hunter-gatherers in general. While subsequent research on watercraft has contributed to a broader comprehension, watercraft remain objects and pragmatic technologies, without further considerations of their other potential dimensions. Usually situated in ecology and political economy frameworks, these approaches have posed epistemological and methodological limitations. In the PNWC, dugout canoes were more than just boats; they were wholly integrated into cultures, cosmologies, and mythologies and were regarded as beings that were born, lived, and died. People traveled, paddled to feasts, procured resources, went to war, explored new territories—all while going by boat-being. Because of this, paddling in boats entailed markedly different human–watercraft relationships, which warrants further consideration in building watercraft theory (Smith 2022). Like Ames' (2002) article, this chapter is about boats, but it is about boat cultures and boats in pluralistic realities.

The rich ethnographic record of the PNWC captured glimpses of human–watercraft relationships in which dugout canoes played key roles in epic, cautionary, and mundane narratives, stories, and performances. However, besides canoe parts found at the wet archaeological site Ozette (45CA24), PNWC precontact canoes have not survived the challenges of time. The Ozette collection consists of several canoe fragments, canoe board parts recycled as house planks, numerous canoe figurines, and other iconographic associations (Bernick 2017; Daugherty 1988; Mauger 1991; Wallace 2017). In addition, canoe bailers, paddles, and anchor stones still tied with cordage have been found at Ozette, Little Qualicum River (DiSc-1), and Lachane (GbTo-33; see Bernick 2017). Nonetheless, PNWC precontact canoes can be considered ghost images (Bjerck 2016); although we may no longer have direct evidence of their presence, they left signatures and were undoubtably essential to livelihood and ways of being.

The concept of a watercraft culture has previously been used to connote how transformative watercraft have been to boat-dependent peoples (e.g., Ames 2002; Des Lauriers 2005; Durham 1960; Neel 1995). Situating transportation, naturalist John Muir once compared the significance of the mustang horse of the Mexican vaquero to the dugout canoe of PNWC peoples (Durham 1960). David Neel (1995) expanded on this by comparing the importance of PNWC dugout canoes to the automobile today. Both of their statements provide context for the revolutionary capabilities of transportation in which taming the horse, building a canoe, and manufacturing a car transformed cultural landscapes.

Furthermore, watercraft can be conceptualized as cultural keystones (akin to ecological keystones) that play key roles in the structuring and functionality of culture—a concept previously explored in the PNWC to address the significant roles organisms such as salmon and cedar hold within ecosystems (Garibaldi and Turner 2004; Patrick 2004). Yet this emphasis has been limited to the organism or resource as a keystone over the cultural material generated. I support a more nuanced understanding of the cultural keystone model in which a dugout itself carried on the essence of the cedar tree and became another active being that not only had an unprecedented level of integration into PNWC cultures but also extraordinarily shaped cultures and ways of being—they informed human and other-than-human worlds, were vessels that could travel between the realms, and were transformative beings with land–sea associations. I demonstrate this by situating dugouts within human lives—both referentially and through homologies, which are ontologically informed resemblances found in ethnographies, art, and archaeology, and other domains.

Today PNWC dugouts are no longer ghost images; importantly, among many peoples, they are part of resurgent watercraft cultures, and dugouts remain cultural keystones holding significant positions in both individual and communal lives. Canoes remain a part of pluralistic identities and have taken on new roles in cultural resurgence and maintenance, in reconnecting peoples to the past and ancient waterways, and as healthy impetuses in youth-focused activities, among serving other vital culture-enriching functions. Canoe Journeys have become essential components in these efforts through which Indigenous peoples learn canoe construction and paddling and celebrate their cultures as part of the PNWC dugout revival. Boats continue to be more than just pragmatic transportation vessels and, to many PNWC peoples today, remain boat-beings.

Northwest Coast Dugouts

Watercraft were ubiquitous in the PNWC, vital for survival and essential in livelihood, but, like horses and cars, they were not homogeneous; rather, several factors influenced the heterogeneity we see in them. For instance, the availability of construction resources, the intended functions and maneuverability requirements of the craft, the types of aquatic environments for use, and construction knowledge were all factors.

PNWC oceangoing watercraft can be conceptually divided into two styles: vernacular and specialized (Smith 2022). Both craft existed on a continuum of sizes. Vernacular watercraft were general-purpose craft serving routine functions and used daily. Construction materials were usually unrestricted and available to most people, and construction knowledge was commonly held. Examples include dugout canoes or tule balsas, which were common along the Pacific coast, among other craft. Specialized watercraft generally served more specific functions requiring certain capabilities relating to speed, stealth, and hauling capacity. Likely more organization, coordination, and people would be necessary for construction, rowing power, and maneuvering tactics (Smith 2022). In specialized craft, construction resources were usually more restricted or selectively used. Often only a limited number of people had the knowledge and skill to build these vessels, and the process could be both time-consuming and expensive. Examples along the Pacific Coast include large dugouts such as the Great Canoe (e.g., Neel 1995), sewn-plank canoes like Chumash *tomols* (e.g., Hudson et al. 1978), large tule balsas (e.g., Smith 2022), and composite craft like Isla Cedros canoes (e.g., Des Lauriers 2005). No watercraft was more "complex" than another; rather, they were all incredible innovations fine-tuned over multiple generations to support the functions and purposes at hand (Smith 2022).

While there have been various attempts to classify PNWC oceangoing canoes, the most concise system identifies three main canoe types: the Westcoast Type, Northern Type, and Coast Salish Type (e.g., Durham 1960; Neel 1995; Stewart 1984; Suttles 1990; see Smith 2022) (Figure 9.1). Each type had a continuum of sizes, specialized and vernacular versions, and local modifications to styles. Specializations often related to activities such as freighting, warring,

Figure 9.1. Three types of PNWC oceangoing canoes (original illustrations based on canoes in the collection of the Canadian Museum of History): (*a*) Northern Type (likely from Queen Charlotte Islands, British Columbia, late nineteenth century, CMC VII-B-1126); (*b*) Westcoast Type (Duncan, British Columbia, 1929, CMC VII-G-346); (*c*) Coast Salish Type (Duncan, British Columbia, 1929, CMC VII-G-352). Canadian Museum of History, Gatineau, Quebec.

whaling, and ceremonialism. Some canoes were traded between groups and used outside of the area in which they were constructed (Durham 1960). Others were specifically traded without decorations, resulting in certain canoe styles decorated with outside iconography (Stewart 1984).

Westcoast Type dugouts were used by Wakashan-speaking groups as well as some Salish-speaking peoples and grew to become one of the most widely used canoes in historical times. The vessels were made by peoples on the west coast of Vancouver Island as well as in Washington and Oregon. Among whaling groups, specialized canoes were locally constructed by the Nuu-chah-nulth, Makah, Chinook, and Quileute (Durham 1960). Northern Type dugouts were used by the Haida, Tsimshian, Bella Bella, Haisla, Tlingit, and Kwakwaka'wakw. For specialized canoes, local modifications were made, such as with the Haida design to refit prows to transform canoes for either dedicated war or ceremonial uses (Durham 1960; Seaburg and Miller 1990). The Coast Salish Type dugout was used by the Northern, Central, Southern, and Southwestern Salish, geographically throughout the Salish Sea, with local variations, including the Puget

Sound and Northern Gulf canoe styles (Holm 1991; Suttles 1987). Salish canoes could be strung together, sometimes catamaran style, to freight heavy loads over distances, like hauls of salmon or house beams during residential moves. Regional variations in dugouts, like houses, paralleled the diversity of social organizations often distinguishing northern, central, and southern PNWC groups (Suttles 1987). Notably, all three types of PNWC dugouts could handle the open ocean, straits, sounds, and any other sort of waters alike, when conditions were favorable and predictable, but when calm waters changed without warning, lives could be in danger regardless of vessel type (Smith 2022).

Boats' Origins and Living Relationships

Watercraft have been alongside peoples in the PNWC since the beginning of time, facilitating access to places otherwise untravellable. To some people, such as the Salish Lushootseed, the cosmos was understood as a capsized canoe, meaning the world was a canoe or existed within a canoe (Miller 1999). Similarly, stories of the Great Floods are common in PNWC cosmologies. Although there are variations, a common version tells of the Myth Age when the floods came, and both supernaturals and ordinary people boarded canoes and survived by floating until the waters receded (e.g., Boas 1916; see also Hutchings and Williams 2020). Where the canoes landed or were anchored became important places as these were the villages from which people descended. Archaeological research further supports long-forged human–watercraft relationships from early British Columbian archaeological sites on Triquet Island, Kilgii Gwaay Island, Calver Island, Quadra Island, and elsewhere (Braje et al. 2020; Fedje et al. 2005; Fedje et al. 2018; Gauvreau and McLaren 2017; Koppel 2016; McLaren et al. 2018; Potter et al. 2017; Potter et al. 2018).

Peoples had both historical and living relationships with canoes (Heidegger 1977; Ingold 1993). Experiencing the world in a canoe created and evoked memories and histories, and it generated wisdom through personal observations, the learning of place names, and narratives tied to places. Canoes were place-making, locales of wisdom through phenomenological experiences (Basso 1996; Heidegger 1977; Tilley 1994). A conscious living relationship emerged between peoples, boats, and the world/s while dwelling in the world/s (Heidegger 1977; Ingold 1993). Boats were places themselves that were meaningfully afloat in space, anchored to places of significance (Smith 2022).

Over the course of human lives, dugouts were coupled with every stage. These living relationships began in infancy and were reaffirmed throughout life, at death, and into the afterlife. Some accounts detail infants placed in dugout-shaped cradles and babies secured to canoe-shaped cradleboards (Durham

1960; Silverstein 1990; Swan 1857; see also Hajda 1990). Children were given miniature wooden dugouts as toys for play, and they learned how to use canoes at an early age, paddling in child-sized boats in shallow coves (Durham 1960). Canoes also played a role in marriage ceremonies: couples were married in canoes, canoes were used as serving bowls during marriage feasts, and canoes were sometimes given as nuptial gifts. The association between marriage and canoes can be further illustrated in the Skidegate expression "to put a string on it," which serves parallel uses for marriage and anchoring a canoe (Swanton 1905: 90; Durham 1960; see below).

Upon death, peoples continued to be attached to boats. They were buried in canoes and together interred into graveyards, placed on beaches, sent out to sea, hung from trees, and added to totems (e.g., Elmendorf 1960; Harper 1971; Holm 1991; Silverstein 1990). Sometimes canoes were "killed" by poking holes through them—the intention was likely to ritually let water pass through as a way to decommission boats and make them unusable by thieves, or as a practice required for access to the afterlife (Miller 1999). In modern times, perhaps one of the most historically known persons interred in a dugout was Chief Seattle, memorialized on a totem in Suquamish, Washington. Canoes were also important in narratives of sickness and in the afterlife. For instance, canoes were incorporated into the Coast Salish spirit canoe ceremony, carrying the shaman into the underworld to cure the living sick (Waterman 1930). Furthermore, some tales of the afterlife included Blue Jay, or a similar being, who travels to the Land of the Dead in a canoe with holes in it (see McKay 2002).

Boat Homologies

Frequent in ethnographic literature are homologies. These are kindred associations between two or more ideas, practices, and things that can be separated by time and space but share similarities in their underlying structure (e.g., Bourdieu 1990; Lévi-Strauss 1976). Homologies shape the strata of consciousness and create pluralistic understandings. Peter Whitridge looked at land–sea homologies in the Arctic, particularly the triad of captain–boat–whaling crew to chief–house–residential household, describing them as meaningful juxtapositions and metaphorical equivalences that "freely exchange properties in the form of conceptual attributes and symbolic associations" (Whitridge 2004: 240). Homologies do not need to be exact or systematically transformative; instead, they can be ontologically equal, historical, particular, local, and transient. Similar to Whitridge (2004), I explore PNWC watercraft homologies as serial homologies, symbolic homologies, homologies of practice, semantic homologies, and homologies of beings (Smith 2022). These categories are not mutually exclusive. An approach based on watercraft homologies is a useful way to look at the relationships between beings, the closeness of ontological relatedness, and prescriptions of transformation, which altogether create a better understanding of pluralistic worlds and realities.

Serial Homology

Some repeated resemblances or multiple homologies can be called a serial homology. An example would be the pentad serial homology of boats–houses–bodies–beings–cosmos. For instance, the shell of a canoe, the walls of a house, the outer skin of the body, the rim of the world, and the essence of being were one and the same, and each contained a source of vitality similarly expressed in the serial homology of an inner soul–heart–hearth–helios (Miller 1999). This similitude often pertains to anatomy, structure, architecture, spirit, and the world. References to house architectural features exemplify this type of serial homology. The fixed, two-pitched gable-roof house is common in Northern and Wakashan areas, consisting of a ridgepole, support beams, and an entranceway, among other features. The horizontal ridgepole was thought of as the spine, river, and Milky Way (Miller 1999). The four support poles of the house were the two front arms and two back legs corresponding with the ridgepole spine, or the four sky pillars corresponding with the ridgepole Milky Way (Miller 1999). George MacDonald (1983) observed the Haida house as a representation of the house–body–being–cosmos. It was the axis of the house that was symbolically and ceremonially significant, connecting the underworld, middle world, and upperworld. The centroid of the axis was the central hearth in the house, smoldering and alive. The smoke would rise through the roof hole to the domain of celestial houses; the hearth would warm the living world, and the coals would singe the perimeter of the underworld.

Correspondingly, a serial homology of the heart–hearth–helios, or a slightly differing rendition relating to smoke-breath, is also commonly attributed to boats. Accounts describe the regular presence of a hole in a functioning canoe, either in the center, prow, or stern. For instance, the Tsimshian consider the center hole as the heart of the canoe; without it, the vessel was believed to be unsafe (Miller 1999). Beyond a boat hole, there are accounts of a central fire or smoldering kindle in a boat, like a house hearth but often attributed to nighttime waterfowl hunting, to warming crews, and to cooking fish while afloat (Elmendorf 1960; Lincoln 1992; Miller 1999; Suttles and Lane 1990). Altogether, the smoke from house and canoe hearths is homologous to breath in beings, like the exhale of a human and the spouting release from the blowhole of an orca.

A further parallel of the triad boat–house–body serial homology is that they also share various bodily orifices. They are present in accounts and early photographs of house entrances, specifically among the Northern area groups. At the Haida village of Skidegate, House 5 had an entrance representing a mouth, and House 31 had an entrance resembling a vagina. Similarly, other Haida houses had entrances through the navel of an animal crest (MacDonald 1983). Some houses had interior division screens to separate the chief's quarters, and these entrances could be bodily orifices. Like houses, canoes also had navels, mouths, and vaginas, among other anatomies. All of these shared attributes indicate houses and canoes as complex organic life—they are all alive.

The homologous relationship between canoes and houses can also be understood as one of material confluence and physical transformation. In a canoe burial, the body and boat decay into one another. Another example comes from Ozette where worn-out canoe board fragments were recycled into wall planks, conjoining the canoe and house as one (Mauger 1991; Wallace 2017). The physical and transformative connection of boat–house is also present in the Coast Salish spirit canoe ceremony, in which a sick person is placed in a room and surrounded by spirit canoe boards and spirit helper beings. The house transforms into a canoe, and the shaman can then travel to the underworld to find the evil spirit causing the sickness (Lincoln 1992; Pasztory 2005; Underhill 1965; Waterman 1930).

Symbolic Homology

Often symbolism and ritual were associated with manufacturing a canoe, akin to creating and birthing the vessel. To many PNWC peoples, such as the Coast Salish, Tsimshian, Tlingit, and others, canoes were living beings (De Laguna 1990; Miller 1999). The guardian spirits, Woodpecker and Cedar, were believed to have encountered certain human people and given them the instruction and ability of canoe-carpentry powers in dreams. The canoe builder and the tree would collaborate during the felling so that the fall would not cause damage to the grain or unnecessary harm to the forest. Once the canoe was carefully carved, the canoe maker and his wife were believed to be the parents (Miller 1999). The spirit of the cedar, perhaps even thousands of years old in old-growth forests, never leaves during the process of canoe making (e.g., Johansen 2012; Sarvis 2003). Among the Tsimshian, there are accounts that once a canoe was created, it would be launched by the parents and christened as a living being (Miller 1999). Certain Coast Salish accounts describe the gender of the canoe according to their grain. For instance, canoes created with coarse-grained wood were *palowqwtan* and considered masculine. Canoes of fine-grained wood, *paloqwatant*, were feminine by nature. The ways in which people spoke to canoes was informed by the gender of the canoe (Lincoln 1992; Miller 1999).

Homology of Practice

Habitual, customary, mundane, and sacred actions can be a homology of practices when they share similitudes of meanings. In action, they connect the mundane to ritual and the secular to sacred, and it becomes the process that is equally meaningful to the outcome. Fundamentally, a homology of practice informs ontological principles applicable to living in the world and maintaining equilibrium (Smith 2022). Following the boat–house–body serial homology, in practice, bailing water from a canoe, cleaning the house, and curing the body of illness were one and the same (Miller 1999). In Makah performance ceremonies, the boat–house–being serial homology is expressed in paddling a canoe, harpooning the house of a bride, and taking a whale (Durham 1960).

Semantic Homology

A homology of words, phrases, names, meanings, and interpretations can be called a semantic homology even if the contextual influences are different. A semantic homology can show connections between words and phrases and the ways in which lexical similitudes of words and phrases become shared meanings, references, and truths. This involves mental phenomena that link similitudes of words and phrases, creating layers of meanings. Peoples' understanding of a semantic homology determines how they comprehend things and ultimately informs their actions. An example is the Skidegate expression above, in which the same term was used for marriage and anchoring a canoe (Durham 1960; Swanton 1905). A canoe–body semantic homology is illustrated in Tsimshian language, in which the word for canoe means water travel but is also colloquially understood as "vehicle, waterway, narrow passage, throat, body trunk, curved sided" (Bates et al. 1994: 103; see also Miller 1999). Some Coast Salish referred to the nob on the neck of the canoe as the "navel" while the Makah called this the uvula. Other references have been previously mentioned as a serial homology, such as references to canoe heart–hearths (see above). Another example concerns naming practices among the Kwakwaka'wakw, Tlingit, and others in which names for canoes, houses, and people were shared and often associated with crests. As such, the inheritance of and access to crests and rights were connected as semantic homologies.

Homology of Being

Part of many non-Western pluralistic ontologies entail living in the world/s and interacting with other-than-human beings, some of which had personhood and transformative capabilities. A homology of being can be human and other-than-human beings that fundamentally share humanity in heart/soul but simply wear different skins/cloaks and can change based on ontological prescript. Importantly, it is the condition of humanity among all beings that is primordial plenum (Viveiros de Castro 1998; 2004). Brazilian anthropologist Eduardo Viveiros de Castro (1998; 2004) offers perspectival multinationalism to help conceptualize these Amerindian ontologies. Viveiros de Castro's method involves multinaturalism, where perception is the point of view located in the body and understood as valid, in contrast to representation, which resides in the mind and spirit; transmorphism, which allows access to multiple bodies; and perspectivism, which accepts multiple viewpoints. For instance, having different points of views is due to having different bodies, as the things seen and experienced are from these particular perspectives. Viveiros de Castro characterizes the perception of the subject in relational systems with his classic relational pointer: what blood is to humans, beer is to jaguars (Viveiros de Castro 2004; see also Smith 2022).

The ontological method of perspectival multinaturalism can be an equally useful approach in the PNWC. Robert Losey (2010) uses perspectivism in research on animated fishing weirs as longhouses for other-than-human fish persons at Willapa Bay, Washington, that mediated the catch-and-kill relationship between fish and humans (Boas 1966; De Laguna 1972; Suttles 1974). Failing to dismantle fish traps off-season was considered neglectful and created unnecessary waste. Fish persons would react by no longer seeing the weirs as usable longhouses, resulting in fish persons subsequently seeking shelter elsewhere and fewer fish for humans to catch.

Similarly, Alan McMillan (2019) uses an ontological approach to effigies and pictographs on western Vancouver Island of orcas and female anthropomorphs. In combination with ethnographies on local contact-era whaling groups (Nuu-chah-nulth, Ditidaht, and Makah), McMillan draws associations between the whaling chief, his wife, and her role in whale hunts. McMillan illustrates the powerful supernatural connection between a whaling chief to whale predators as other-than-human whalers (e.g., thunderbird, wolf, orca, supernatural lightning serpent). Ontological associations may have deep historical roots, perhaps over one thousand years. The whaling house at Ozette included several thunderbird objects and an orca sculpture/saddle—considering the context, both seem to be elements of the whaler–thunderbird–wolf–orca–serpent homology of beings.

The connection between orcas, wolves, and canoes is commonly portrayed as homologous beings in ethnographies. The orca–wolf–human homology of being is a land–sea counterpart: they all live in packs, practice cooperative hunting, and share carnivorous diets. An orca in the water transforms into a wolf on

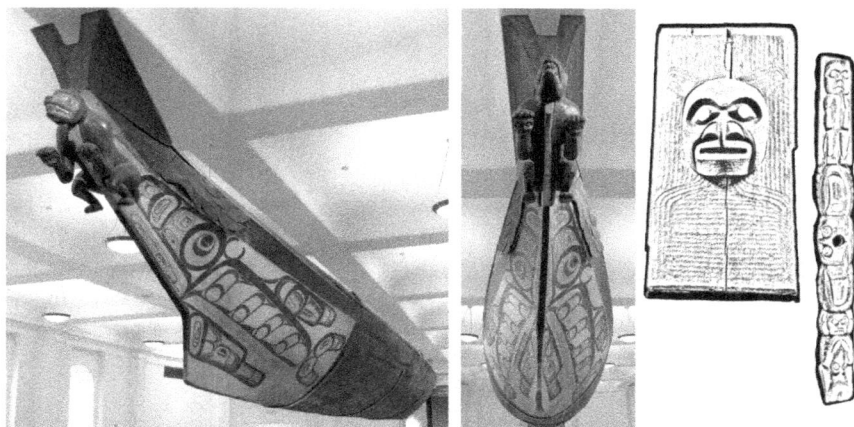

Figure 9.2. Great Canoe, 63 ft., created by Haíłzaqv (Heiltsuk) and Haida artists in 1878. Although its given name has been lost, it is simply called canoe—*ǧḷ'w'a* in Haíłzaqvḷa (Heiltsuk), and *tluu* in Xaayda Kil (Haida). The canoe has a carved and painted prow with wolf crouched and baring fangs, and a hull with orca in geometric formation. Inside the canoe are carved anthropomorphic benches. The interior stern has a high-relief human face protruding from a flat but textured anthropomorphic background carving. *Left and Middle*: Wikimedia Commons, photos by Tony Hisgett. Licensed under CC BY-2-0, https://creativecommons.org/licenses/by/2-0/. *Right*: Drawing by Alberto García-Piquer.

land, with the dorsal fin turning into the wolf's tail and vice versa (see McMillan 2019). A similar embodiment of the orca–wolf–canoe homology of being can be seen in Nuu-chah-nulth whaling canoes that were painted black, as orcas in disguise, serving to hide them from prey. Secret animals, like the wolf, were carved on the canoe's prow (Miller 1999). Homological associations between human and other-than-human whaler beings have also been more conspicuously represented in ornately painted and carved historical and contemporary canoes, with some exhibiting quite complex relational imagery (Figure 9.2).

An orca–canoe homology of beings demonstrating transformational agency can be found in various stories and myths. The Haida story of Skana (Orca) tells of a man and canoe that transform into the spirit of the man inside an orca. One version was recorded by ethnographer James G. Swan during his 1883 Smithsonian-funded expedition to Haida Gwaii and was illustrated by the Haida artist and interpreter Johnny Kit Elswa (Elswa 1883; Judson 1911; Niblack 1888; Figure 9.3). Similarly, in Nuu-chah-nulth stories of planned interactions between orcas and humans, an orca person in human form residing in an underwater longhouse boards a canoe to head to shore. Transmorphism

Figure 9.3. *Skana the Killer (Orca), Haida Mythology.* Pen and ink drawing by Johnny Kit Elswa (1883). The depiction also appears as "The man-spirit was inside the Skana" (Judson 1911) and is followed by a Haida story about killer whale. Men in a canoe came across an orca and they threw rocks at its fins. The orca goes to the nearby beach. Later the men see smoke on the beach but only find a man cooking dinner near his canoe. The man asks why they threw rocks at him and then shows them a transformation from a man in a canoe to the man's spirit inside an orca. Franz R. and Kathryn M. Stenzel Collection of Western American Art. Yale Collection of Western Americana, Beinecke Rare Book and Manuscript Library.

then occurs: the canoe becomes an orca body, and the orca person's spirit relocates to the dorsal fin (Golla 2000; Sapir and Swadesh 1939; Sapir et al. 2004; see also McMillan 2019). The transformation embodied in this homology of beings is materialized in a carved cedar orca dorsal fin sculpture/saddle uncovered in a whaling house at Ozette and elaborated with sea otter teeth outlining a thunderbird and serpent (McMillan 2019). In 1778 John Webber created a historical illustration depicting Yuquot Village, which similarly shows a dorsal fin sculpture/saddle displayed in a house (Marshall 2000; see also McMillan 2019). Both are likely portrayals of a homology of beings that connect human and other-than-human whalers, rather than simply trophies representing hunting prowess and chiefly power. Accounting for sexual dimorphism in orcas and the shape of the dorsal fins, both renderings likely indicate female orcas (Ford 2014; Naughton 2012; see also McMillan 2019)—a topic worth exploring further as part of these ontological relationships.

Post-Contact Domestication of Canoes

As with peoples on the PNWC, canoes and canoe-beings persisted through the challenges of historical times and continue to thrive today. However, colonization resulted in major disruptions in watercraft building and use, which had fundamental repercussions to watercraft cultures along the Pacific coast and elsewhere. Along with peoples, Indigenous watercraft were also controlled, appropriated, colonized, assimilated, and stolen (Ritts et al. 2018). Recognizing watercraft as a cultural keystone, targeting canoes was a skillful tactic to degrade cultural cohesion, facilitate colonial acculturation, and disconnect Indigenous pluralistic ontologies.

Indigenous canoes were used by colonizers who hired Indigenous peoples for their extraregional knowledge and skills to paddle, navigate, and guide expeditions along rivers into the continent's interior, where colonizers assessed the profitability of some of the natural commodities within these "new" landscapes. Max Ritts and colleagues (2018) emphasize how Indigenous canoes became the vehicles in which colonizers encountered Indigenous peoples, and through ongoing encounters and interactions, canoes became part of an unspoken infrastructure of colonial expansion.

Perhaps the most well-known example of canoe colonization was the US government–funded exploration by Meriwether Lewis and William Clark, who hired Indigenous guides to navigate the complex continental river systems from St. Charles, Missouri, to the Columbia River and Oregon coast, and back. After failed attempts to exchange for a canoe with the Clatsop Chinookans, they had four men in their expedition party steal a canoe (Erickson and Krotz 2021). Stealing a canoe was kidnapping and enslaving a being without consent, and the perpetrators were without any critical knowledge on how to properly take care of the vessel-being. The wrong was eventually ameliorated by Clark's descendants 200 years later, who commissioned and gifted a new canoe back to the Chinook, who welcomed it (Erickson and Krotz 2021).

Canoe-integrated cultural practices were greatly diminished by Canada's Indian Act of 1876. Land was taken away, people were moved, land was privatized, and the lumber industry intensified, all separating people from trees for canoe building (Cushman et al. 2021). Ceremonies and gatherings, such as the potlatch, were banned, which disconnected canoes from these meaningful practices and decreased the need to use canoes (Cushman et al. 2021). And canoes became appropriated, modernized, and standardized—perceived as improvements—and used by settlers as industrial, company-owned fleets (Ritts et al. 2018). Some canoes wore the company stamp—domesticated, owned, branded like cattle. Ritts and colleagues (2018) interpret historical photographs of company canoes positioned in front of traditional longhouses as representations of the domestication of Indigenous labor. But throughout this, peoples and canoes persisted. Due to cultural resilience and the proficiency of canoes to new industries and the fur trade, remarkably, there were approximately 10,000 canoes on the British Columbia coast at the beginning of the twentieth century (Ritts et al. 2018).

Contemporary Canoes Shape Identities and Cultural Landscapes

PNWC peoples and their dugouts helped shape the modern world, and the canoe has become a significant national symbol of Canada. In fact, in 2007 the Canadian Broadcasting Corporation announced the canoe at the top of the list of the Seven Wonders of Canada out of over 25,000 nominations, including the Rocky Mountains, Northwest Passage, CN Tower, and the Montreal-style bagel (Osler 2014). Commentator Peter Mansbridge reflected: "It's hard to imagine Canada being Canada without the canoe. Explorers, missionaries, fur traders and First Nations—they're all linked by this subtle and simple craft. To many, the quintessential Canadian experience begins by picking up a paddle. That's why the canoe is one of the seven wonders" (Osler 2014: 12). However, in the contemporary cultural milieu of canoe admiration, it is also important for the public to recognize the history of canoe appropriation, Indigenization, and fetishization (Dean 2013). By doing so, everyone can be part of the process in understanding the complexities of the relationships between peoples and canoes. Misao Dean (2013) argues that through canoe appropriation, Westerners have used canoes to tie themselves to First Nations peoples in an unbroken line of descent, which then allows them to view themselves as the rightful inheritors of the land.

Watercraft among First Nations and Native Americans today are part of movements centered on decolonization and sovereignty, Indigenous resurgence, colonial reconciliation, cultural revitalization, community heritage and healing, and the enrichment of tribal identity. Canoe Journeys, paddling workshops, and canoe-building workshops are enhancing communal Indigenous identities and healing intergenerational traumas from assimilation (Brown et al. 2021). One of the first Canoe Journeys was the 1989 Paddle to Seattle hosted by the Quinault that included 17 Tribes from Washington state; the event celebrated the one hundredth anniversary of Washington statehood and the recognition of Indigenous sovereignty in the Centennial Accord. Today there are a handful of Canoe Journeys. Paddle to Seattle carried on, subsequently hosted by various tribes and organized to travel between different locations each year; the most recent, the 2023 Paddle to Muckleshoot, was hosted by the Muckleshoot Tribe.

Canoe Journeys and workshops have also been oriented toward tribal youth development, designed to instill cultural values and strengthen communal ties while also focusing on healthy lifestyles and suicide prevention. This is a significant contribution in a time when Indigenous peoples are being lost to suicide at exponential rates (e.g., Stevenson 2014; Oh 2016). Paddling in dugouts reconnects peoples to cultural heritage, strengthens community, and reunites peoples with ancestral lands and ancient water pathways. Children and teens learning the history of canoes and traditional methods to paddling are not only enculturated but also instilled with a powerful sense of identity (Oh 2016). Engagement with canoes and learning watercraft culture in the PNWC prevent children from losing their cultural identities. Canoes save lives—from the Myth Age Great Floods to today.

Persisting Indigenous Canoe Ontologies

The revival of watercraft culture also entails renewing traditional relationships between humans and watercraft. This includes an emphasis on prayers during felling, communication between human and wood, recognition of individual canoe spirits, awareness of the sacredness of travel through time and realms, and the importance of personal purity (see Sarvis 2003). Makah elder Mary T. Greene McQuillen (Kwe-de-che-autlh), spoke of how canoes have cedar spirits and feel everything (Neel 1995; Sarvis 2003), so McQuillen sang, danced, and greeted canoes with unwavering respect, notably performing the Thunderbird dance in front of an "ancient" canoe at Point Hudson over 50 years ago (The Leader 2019). This example shows the persisting ontology of a homology of beings among human and other-than-human whalers and canoes. PNWC canoes have always been beings with personhood and selfhood, as individuals with unique personalities. In the 1980s Haida carver Bill Reid (Iljuwas Kihlguulins Yaahl SGwansung); Haida carver, singer, and tribal leader Gary Edenshaw (Gidansda Giindajin Haawasti Guujaaw); and Kwakwaka'wakw carver Simon Dick (Tanis), among others, researched and built the 50-ft. canoe *Loo Taas* (Wave Eater), spirited with a 750-year-old cedar log. Edenshaw spoke about the human commitment to a canoe—the vessel requires your time, needs to be properly cared for, and can get jealous if not given your undivided attention (Johansen 2012).

Figure 9.4. Image still of birch bark canoes in a prison from the short film, *How to Steal a Canoe* (Simpson and Strong 2016), based on the poem by Leanne Betasamosake Simpson (2017). Canoe-beings continue to be part of Indigenous ontologies today. Transgenerational colonial traumas are not limited to Indigenous humans, but also extend to other-than-humans alike, and this suffering is apparent with canoes in museums, among other-beings narrowly considered "cultural relics." Through various artistic mediums, wider audiences are being reached, challenging people to think differently about canoes from outside the confines of Western worldviews. Spotted Fawn Productions, Vancouver, British Columbia, CA.

Contemporary Indigenous voices echo canoe homologies from ethnographies, and canoe-beings are part of contemporary Indigenous resurgence. In a provocative portrayal of canoe-beings, Anishinaabe poet Leanne Betasamosake Simpson's (2017) poem and clay film *How to Steal a Canoe* (Simpson and Strong 2016) address the theft of birch bark canoe-beings and their imprisonment in museums (see Erickson and Krotz 2021; see Peterson 2020; Figure 9.4). Specifically, Simpson accuses museums of participating in settler colonialism and, in effect, preservation violence. Canoes hanging in museums and collection warehouses are homologous to canoe-beings hung in jails (Smith 2022).

Context and consent are of particular relevance. There are consensual arrangements in which canoe-beings and their families have agreed on the placement of canoes in museums, either for safekeeping or to proudly display as part of storytelling and sharing traditions. However, there are other cases in which consent has not been given. Some canoes were stolen for study or popular exhibition. Simpson equates stolen canoes as theft of bodies, and states "akiwenzii [old man] say, 'oh you're so proud of your collection of ndns [*sic*]. good job zhaaganash [white man], good job'" (Simpson 2017: 69). Without proper ritual or care requirements, stolen and imprisoned canoe-beings were abused, tortured, and angry. Simpson (2017: 69) describes the conditions and experiences of imprisoned canoe-beings as "stolen canoes / bruised bodies / dry skin / hurt ribs / dehydrated rage."

Stl'atl'imx Interior Salish storyteller Peter Cole (2006) presents a Canoe Journey shaped out of words and shares a story told along a canoe. In the poem "Relics," Cole (2006: 13) gives a personal account about grandfather's cedar dugout, hung in a museum unable to live its life, unable to naturally decompose, "trapped in time and space / imported time / white space." Cole's narrative further details grandfather's canoe as a war canoe-being, stating, "I look in awe / grandfather / at this tree you have transformed / transfigured / from art of nature / to art of mind / formation / body formulation" (Cole 2006: 13). Cole continues by asking if the canoe will ever again experience the stimulation of all the senses afloat at sea; the feeling and sound of the waves, the touch of the sun, and the smell of the air and mist—the phenomenological realities of boat-beings.

Although these works are about canoe-beings in confinement, the approaches differ slightly. As possible solutions, Simpson focuses on stealing and reclaiming, and Cole offers either setting it free or ending its misery with an axe (Cole 2006; Simpson 2017; Simpson and Strong 2016; see also Erickson and Krotz 2021). Both Indigenous artists detail experiential suffering of canoe-beings, rotting and wasting away in jails without being able to experience life afloat. The narratives punctuate the contemporary realities of canoe-beings and the hope to finally be released to experience life or death in natal waters.

Conclusion

It has been over 20 years since Ames' (2002) considerable contribution to watercraft scholarship and over 50 years since the beginning of the Canoe Movement in Hawaii with the double-hull canoe *Hokule'a*. Watercraft research and theory need greater expansion outside of and in addition to traditional political economy and ecological frameworks. In this chapter I have focused on an Indigenous ontological approach and explored the relationships between humans and watercraft in cosmology, as living relationships through stages of life and death, as homologous associations, and as beings. In the PNWC, watercraft have primordial connections with humans, are transformative beings with humanity as their basis, and have shaped cultures and consciousnesses. The dugout is part of the modern world as a symbol of the collective and pluralistic identities of both Indigenous and non-Indigenous peoples.

Although precontact Northwest coast watercraft may be ghost images in archaeology, watercraft today are overtly visible and integrated into resurgence cultures. PNWC peoples continue to produce watercraft cultures that help structure, inform, and guide ways of being. And watercraft are the vehicles that help navigate through the intersectionality of Indigenous and Western realities. Continued watercraft research, new theoretical applications, and Indigenous collaboration on watercraft theory-building can be invaluable in gaining further insight into the archaeological past, reconnecting the past to today, and instilling a deeper sense of cultures. Going by boat-being shares the same waters as going by boat, but the boat-beings help steer along a different course through the overlapping realms of watery worlds and pluralistic realities.

Acknowledgments

I would like to thank the editors for inviting me to contribute to this volume and for putting together such a diverse collection of research on watercraft and their uses. Thanks also to the reviewers for their useful comments. And a special thanks to Colin Grier, Andrew Duff, and William Andrefsky Jr. for reading and commenting on an earlier version of this manuscript from my dissertation at Washington State University.

References Cited

Ames, Kenneth M. 2002. Going by Boat: The Forager-Collector Continuum at Sea. In *Beyond Foraging and Collecting: Evolutionary Change in Hunter-Gatherer Settlement Systems,* edited by B. Fitzhugh and J. Habu, pp. 19–52. Plenum, New York.

Basso, Keith H. 1996. *Wisdom Sits in Places: Landscape and Language among the Western Apache.* University of New Mexico Press, Albuquerque.

Bates, Dawn E., Thom Hess, and Vi Hilbert. 1994. *Lushootseed Dictionary.* University of Washington Press, Seattle.

Bernick, Kathryn. 2017. Archaeological Remains of Precontact Watercraft on the Northwest Coast. *Alaska Journal of Anthropology* 15(1–2): 49–65.

Bjerck, Hein B. 2016. Settlements and Seafaring: Reflections on the Integration of Boats and Settlements among Marine Foragers in Early Mesolithic Norway and the Yámana of Tierra del Fuego. *Journal of Island and Coastal Archaeology* 12(2): 276–299.

Boas, Franz. 1916. *Tsimshian Mythology.* Based on texts recorded by Henry W. Tate, Thirty-First Annual Report of the Bureau of American Ethnology, 1909–1910, pp. 27–1037. Government Printing Office, Washington, D.C.

Boas, Franz. 1966. *Kwakiutl Ethnography.* University of Chicago Press, Chicago.

Bourdieu, Pierre. 1990. *The Logic of Practice.* Translated by R. Nice. Stanford University Press, Stanford, California.

Braje, Todd J., Jon M. Erlandson, Torben C. Rick, Loren Davis, Tom Dillehay, Daryl W. Fedje, Duane Froese, et al. 2020. Fladmark + 40: What Have We Learned about a Potential Pacific Coast Peopling of the Americas? *American Antiquity* 85(1): 1–21.

Brown, Frank, Hillary Beattie, Vina Brown, and Ian Mauro. 2021. Tribal Canoe Journeys and Indigenous Cultural Resurgence: A Story from the Heiltsuk Nation. In *The Politics of the Canoe,* edited by Bruce Erickson and Sarah Wylie Krotz, pp. 27–47. University of Manitoba Press, Winnipeg.

Cole, Peter. 2006. *Coyote and Raven Go Canoeing: Coming Home to the Village.* McGill-Queen's University Press, Montreal.

Cushman, Rachel L., Jon D. Daehnke, and Tony A. Johnson. 2021. This Is What Makes Us Strong: Canoe Revitalization, Reciprocal Heritage, and the Chinook Indian Nation. In *The Politics of the Canoe,* edited by Bruce Erickson and Sarah Wylie Krotz, pp. 49–71. University of Manitoba Press, Winnipeg.

Daugherty, Richard D. 1988. Problems and Responsibilities in the Excavation of Wet Sites. In *Wet Site Archaeology,* edited by Barbara A. Purdy, pp. 15–29. Telford Press, Caldwell, New Jersey.

De Laguna, Frederica. 1972. *Under Mount Saint Elias: The History and Culture of the Yakutat* Tlingit. Parts One, Two and Three. Smithsonian Institution Press, Washington, D.C.

De Laguna, Frederica. 1990. Tlingit. In *The Northwest Coast,* edited by Wayne Suttles, pp. 203–228. Vol. 7 of *Handbook of North American Indians,* William C. Sturtevant, general editor. Smithsonian Institution, Washington, D.C.

Dean, Misao. 2013. *Inheriting a Canoe Paddle: The Canoe in Discourses of English–Canadian Nationalism.* University of Toronto Press, Toronto.

Des Lauriers, Matthew R. 2005. The Watercraft of Isla Cedros, Baja California: Variability and Capabilities of Indigenous Seafaring Technology along the Pacific Coast of North America. *American Antiquity* 70(2): 342–360.

Durham, Bill. 1960. *Canoes and Kayaks of Western America.* Copper Canoe, Seattle, Washington.

Elmendorf, William W. 1960. *The Structure of Twana Culture.* Washington State University Research Studies, Monographic Supplement No. 2. Washington State University Press, Pullman.

Elswa, Johnny Kit. 1883. *Skana the Killer (Orca), Haida Mythology.* Pen and ink drawing. In Series II: James G. Swan Artwork and Associated Material, 1852–1911. Franz R. and Kathryn M. Stenzel Collection of Western American Art. Yale University Library, New Haven, Connecticut. https://archives.yale.edu/repositories/11/archival_objects/459634.

Erickson, Bruce, and Sarah Wylie Krotz. 2021. Introduction. In *The Politics of the Canoe,* edited by Bruce Erickson and Sarah Wylie Krotz, pp. 1–23. University of Manitoba Press, Winnipeg.

Fedje, Daryl W., Alexander P. Mackie, Rebecca J. Wigen, Quentin Mackie, and Cynthia Lake. 2005. Kilgii Gwaay: An Early Maritime Site in the South of Haida Gwaii. In *Haida Gwaii: Human History and Environment from the Time of Loon to the Time of the Iron People,* edited by Daryl W. Fedje and Rolf W. Mathewes, pp. 187–203. University of British Columbia Press, Vancouver.

Fedje, Daryl, Duncan McLaren, Thomas S. James, Quentin Mackie, Nicole F. Smith, John R. Southon, and Alexander P. Mackie. 2018. A Revised Sea Level History for the Northern Strait of Georgia, British Columbia, Canada. *Quaternary Science Reviews* 192: 300–316.

Ford, John K.B. 2014. *Marine Mammals of British Columbia.* The Mammals of British Columbia, Royal BC Museum Handbook, Vol. 6. Royal British Columbia Museum, Victoria.

Garibaldi, Ann, and Nancy Turner. 2004. Cultural Keystone Species: Implications for Ecological Conservation and Restoration. *Ecology and Society* 9(3), art. 1. http://dx.doi .org/10.5751/ES-00669-090301.

Gauvreau, Alisha, and Duncan McLaren. 2017. Long-Term Cultural Landscape Development at (EkTb-9) Triquet Island, BC, Canada. Paper presented at the 50th Annual Meeting of the Canadian Archaeological Association, May 10–13, Ottawa, Canada.

Golla, Susan. 2000. Legendary History of the Tsisha?ath: A Working Translation. In Nuu-chah-nulth *Voices, Histories, Objects and Journeys,* edited by Alan Hoover, pp. 133–171. Royal British Columbia Museum, Victoria.

Hajda, Yvonne. 1990. Southwestern Coast Salish. In *The Northwest Coast,* edited by Wayne Suttles, pp. 503–517. Vol. 7 of *Handbook of North American Indians,* William C. Sturtevant, general editor. Smithsonian Institution, Washington, D.C.

Harper, J. Russell (editor). 1971. *Paul Kane's Frontier, Including Wanderings of an Artist among the Indians of North America.* University of Texas, Austin.

Heidegger, Martin. 1977. Building, Dwelling, Thinking. In *Basic Writings: From "Being and Time" (1927) to "The Task of Thinking" (1964),* edited by David Farrell Krell, pp. 319–339. HarperSan Francisco, San Francisco.

Holm, Bill. 1991. Historical Salish Canoes. In *A Time of Gathering: Native Heritage in Washington State,* edited by Robin K. Wright, pp. 238–247. Burke Museum and University of Washington Press, Seattle.

Hudson, Travis, Janice Timbrook, and Melissa Rempe. 1978. *Tomol: Chumash Watercraft as Described in the Ethnographic Notes of John P. Harrington.* Anthropological Papers 9. Ballena, Socorro, New Mexico.

Hutchings, Richard M., and Scott Williams. 2020. Salish Sea Islands Archaeology and Precontact History. *Journal of Northwest Anthropology* 54(1): 22–61.

Ingold, Tim. 1993. The Temporality of the Landscape. *World Archaeology* (Conceptions of Time and Ancient Society) 25(2): 152–174.

Johansen, Bruce E. 2012. Canoe Journeys and Cultural Revival. *American Indian Culture and Research Journal* 36(2): 131–141.

Judson, Katharine B. (editor). 1911. *Myths and Legends of Alaska.* A. C. McClurg, Chicago.

Koppel, Tom. 2016. Stepping into the Past. *American Archaeology Magazine* 20(3): 22–29.

Lévi-Strauss, Claude. 1976. *Structural Anthropology.* Vol. 2. Basic Books, New York.

Lincoln, Leslie. 1992. *Cedar Canoes of the Coast Salish Indians.* Center for Wooden Boats, Seattle.

Losey, Robert. 2010. Animism as a Means of Exploring Archaeological Fishing Structures on Willapa Bay, Washington, USA. *Cambridge Archaeological Journal* 20(1): 17–32.

MacDonald, George F. 1983. *Haida Monumental Art: Villages of the Queen Charlotte Islands.* University of British Columbia Press, Vancouver.

Marshall, Yvonne. 2000. The Changing Art and Architecture of Potlatch Houses at Yuquot. In *Nuu-chah-nulth Voices, Histories, Objects and Journeys,* edited by Alan Hoover, pp. 107–130. Royal British Columbia Museum, Victoria.

Mauger, Jeffrey E. 1991. Shed Roof Houses at Ozette and in a Regional Perspective. In *Ozette Archaeological Project Research Reports: 1, House Structure and Floor Midden,* edited by Stephan R. Samuels, pp. 29–173. WSU Department of Anthropology Reports of Investigations 63. Department of Anthropology, Washington State University, Pullman. National Park Service, Pacific Northwest Regional Office, Seattle, Washington.

McKay, Kathryn. 2002. Recycling the Soul: Death and the Continuity of Life in the Coast Salish Burial Practices. MA thesis, Department of History, University of Victoria, Canada.

McLaren, Duncan, Daryl Fedje, Angela Dyck, Quentin Mackie, Alisha Gauvreau, and Jenny Cohen. 2018. Terminal Pleistocene Epoch Human Footprints from the Pacific Coast of Canada. *PLOS One* 13(3): e0193522.

McMillan, Alan D. 2019. Non-Human Whalers in Nuu-chah-nulth Art and Ritual: Reappraising Orca in Archaeological Context. *Cambridge Archaeological Journal* 29(2): 309–326.

Miller, Jay. 1999. *Lushootseed Culture and the Shamanic Odyssey: An Anchored Radiance.* University of Nebraska Press, Lincoln.

Naughton, Donna. 2012. *The Natural History of Canadian Mammals.* Canadian Museum of Nature / University of Toronto Press, Ontario.

Neel, David. 1995. *The Great Canoes: Reviving a Northwest Coast Tradition.* University of Washington Press, Seattle.

Niblack, Albert. 1888. The Coast Indians of Southern Alaska and Northern British Columbia. In *Report of the United States National Museum for the Year Ending June 30, 1888.* Part 2 of the *Annual Report of the Board of Regents of the Smithsonian Institution,* 225–386. U.S. Government Printing Office, Washington, D.C.

Oh, Leslie Hsu. 2016. How Canoes Are Saving Lives and Restoring Spirit. *Smithsonian Magazine,* January 6. https://www.smithsonianmag.com/smithsonian-institution/how-canoes-are-saving-livesand-restoring-spirit-180957712/.

Osler, Sanford. 2014. *Canoe Crossings: Understanding the Craft That Helped Shape British Columbia.* Heritage House, Victoria, British Columbia.

Pasztory, Esther. 2005. *Thinking Through Things.* The University of Texas Press, Austin.

Patrick, Lyana Marie. 2004. Storytelling in the Fourth World: Explorations in Meaning of Place and Tla'amin Resistance to Dispossession. MA thesis, Department of Human and Social Development, University of Victoria, Canada.

Peterson, Max. 2020. Preservation as Violence: The Problem Facing Museums. OHMA blog. Columbia Center for Oral History Research and Interdisciplinary Center for Innovative Theory and Empirics, December 16. http://oralhistory.columbia.edu/blog-posts/People/preservation-as-violence-the-problem-facing-museum-collections.

Potter, Ben A., James F. Baichtal, Alwynne B. Beaudoin, Lars Fehren-Schmitz, C. Vance Haynes, Vance T. Holliday, Charles E. Holmes, et al. 2018. Current Evidence Allows Multiple Models for the Peopling of the Americas. *Science Advances* 4(8): 5473.

Potter, Ben A., Joshua D. Reuther, Vance T. Holliday, Charles E. Holmes, D. Shane Miller, and Nicholas Schmuck. 2017. Early Colonization of Beringia and Northern North America: Chronology, Routes, and Adaptive Strategies. *Quaternary International* 444: 36–55.

Ritts, Max, Kelsey Johnson, and Jonathan Peyton. 2018. Canoes, Modernity, and the Colonial Imagining of Progress. *GeoHumanities* 4(2): 481–503.

Sapir, Eduard, and Morris Swadesh. 1939. *Nootka Texts: Tales and Ethnological Narratives.* Linguistic Society of America, University of Pennsylvania, Philadelphia.

Sapir, Eduard, Morris Swadesh, Alexander Thomas, John Thomas, and Frank Williams. 2004. *The Whaling Indians: West Coast Legends and Stories: Legendary Hunters—Part 9 of the Sapir-Thomas Nootka Texts.* Canadian Ethnology Service Mercury Paper 139. Canadian Museum of Civilization, Gatineau, Quebec.

Sarvis, Will. 2003. Deeply Embedded: Canoes as an Enduring Manifestation of Spiritualism and Communalism among the Coast Salish. *Journal of the West* 42(4): 74–80.

Seaburg, William R., and Jay Miller. 1990. Tillamook. In *The Northwest Coast,* edited by Wayne Suttles, pp. 560–567. Vol. 7 of *Handbook of North American Indians,* William C. Sturtevant, general editor. Smithsonian Institution, Washington, D.C.

Silverstein, Michael. 1990. Chinooks of the Lower Columbia. In *The Northwest Coast,* edited by Wayne Suttles, pp. 533–546. Vol. 7 of Handbook of North American Indians, William C. Sturtevant, general editor. Smithsonian Institution, Washington, D.C.

Simpson, Leanne Betasamosake. 2017. How to Steal a Canoe. In *The Accident of Being Lost: Songs and Stories,* pp. 69–70. House of Anansi Press, Toronto.

Simpson, Leanne Betasamosake (writer), and Amanda Strong (director). 2016. *How to Steal a Canoe.* Spotted Fawn Productions. Vancouver, British Columbia. Film. https://www.spottedfawnproductions.com/how-to-steal-a-canoe/.

Smith, Erin M. 2022. A Multitheoretical Comparative Analysis of Social Organizations and Interaction in California, the Pacific Northwest, and the American Southwest. Ph.D. dissertation, Department of Anthropology, Washington State University, Pullman.

Stevenson, Lisa. 2014. *Life Beside Itself: Imagining Care in the Canadian Arctic.* University of California Press, Oakland.

Stewart, Hilary. 1984. *Cedar: Tree of Life to the Northwest Coast Indians.* Douglas and McIntyre, Vancouver, Canada.

Suttles, Wayne. 1974. *The Economic Life of the Coast Salish of Haro and Rosario Straits.* Garland, New York.

Suttles, Wayne. 1987. *Salish Essays.* University of Washington Press, Seattle.

Suttles, Wayne. 1990. Central Coast Salish. In *The Northwest Coast,* edited by Wayne Suttles, pp. 453–475. Vol. 7 of Handbook of North American Indians, William C. Sturtevant, general editor. Smithsonian Institution, Washington, D.C.

Suttles, Wayne, and Barbara Lane. 1990. Southern Coast Salish. In *The Northwest Coast,* edited by Wayne Suttles, pp. 485–502. Vol. 7 of Handbook of North American Indians, William C. Sturtevant, general editor. Smithsonian Institution, Washington, D.C.

Swan, James G. 1857. *The Northwest Coast; Or, Three Years' Residence in Washington Territory.* Harper and Brothers, and Franklin Square, New York.

Swanton, John Reed. 1905. *Contributions to the Ethnology of the Haida.* (Memoirs of the American Museum of Natural History 8, Part 1.) American Museum of Natural History, New York.

The Leader. 2019. In Memoriam, Mary and the Makahs. The Port Townsend Leader, March 4. https://www.ptleader.com/stories/in-memoriam-mary-and-the-makahs ,60148.

Tilley, Christopher. 1994. *A Phenomenology of Landscape: Places, Paths and Monuments.* Berg, Oxford.

Underhill, Ruth Murray. 1965. *Red Man's Religion: Beliefs and Practices of the Indians North of Mexico.* University of Chicago Press, Chicago.

Viveiros de Castro, Eduardo B. 1998. Cosmological Deixis and Amerindian Perspectivism. *Journal of the Royal Anthropological Institute* 4(3): 469–488.

Viveiros de Castro, Eduardo. B. 2004. Exchanging Perspectives: The Transformation of Objects into Subjects in Amerindian Ontologies. *Common Knowledge (Symposium: Talking Peace with Gods, Part 1)* 10(3): 463–484.

Wallace, Christina L. 2017. *Architecture of the Salish Sea Tribes of the Pacific Northwest: Shed Roof Plank Houses.* James Marston Fitch Charitable Foundation, New York.

Waterman, Thomas Talbot. 1930. The Paraphernalia of the Duwamish "Spirit-Canoe" Ceremony. *Indian Notes* 7(2): 129–148; 7(3): 295–312; 7(4): 535–561. Museum of the American Indian, Heye Foundation, New York.

Whitridge, Peter. 2004. Landscapes, Houses, Bodies, Things: "Place" and the Archaeology of Inuit Imaginaries. *Journal of Archaeological Method and Theory* 11(2): 213–250.

10

Precontact Inuit Watercraft and the Hunter–Prey Actantial Hinge

PETER WHITRIDGE

Watercraft are pivotal components of northern harvesting technologies, enabling the pursuit of game on, beneath, and above the surface as well as travel across watery barriers to access other shores, vastly expanding the economic possibilities and habitable range of coastal peoples. Boats insert themselves between hunters and their prey; walruses congregated at an offshore rookery or pods of bowheads cruising along the coast represent vast but inaccessible stores of resources for hunters without the watercraft, gear, logistics, and tactics to pursue them. The disentanglement of this sociotechnical dilemma is the narrative germ of Eastern Arctic prehistory, and the genius of traditional Inuit technology its solution: specialized watercraft hulled with animal skins, technical clothing and equipment to operate from them, and the organizational skills and ecological knowledge germane to collaborative maritime harvesting in a cold environment. A kayaker armed with harpoon, dart, bow, or lance is among the most iconic images of Inuit in the earliest European depictions (e.g., Sturtevant and Quinn 1989), presumably on account of the technological novelty and practical efficacy of the setup. As explored here, a kayaker pursuing game is also among the earliest motifs in Iñupiat and Inuit figurative engravings. Based on this imagery, alongside the whaling and sealing enabled by boats, certain watercraft appear to have become essential for harvesting the single most important species of terrestrial game—caribou—in at least some settings.

Groups ancestral to Inuit colonized the Western Arctic shorelines of the Bering and Chukchi Seas in the mid-first millennium CE and gradually expanded east onto the Beaufort Sea coast. Beginning around 1200 CE some of these groups rapidly pushed much further east, fanning out across more than 200,000 km of Eastern Arctic island and mainland coast within a couple of centuries and displacing the descendants of Arctic Small Tool tradition groups who had colonized this geographic zone some 3,500 years earlier (see Friesen and Mason [2016] for recent overviews of arctic prehistory). Although a few Eastern Arctic groups eventually came to spend much of their time in the interior (notably, Kivallirmiut bands of the Barren Grounds west of Hudson Bay), the majority occupied coastal settlements during much of their annual round and devoted substantial effort to harvesting marine prey (various species of whale and pinniped) from watercraft. Small watercraft even figured prominently in economically vital caribou hunts on lakes, rivers, and coastal embayments. The patchiness of arctic resources meant that people had to seasonally reposition themselves to effectively harvest game and fish, so mobility was essential for most northern groups; this too frequently entailed the use of watercraft. Alongside breathing hole sealing that transpired on the frozen ocean surface, maritime harvesting from watercraft was central to precontact Inuit lifeways throughout the Eastern Arctic.

Boats were everywhere essential to work, travel, and trade during the open-water season and, although the material record of Inuit watercraft is relatively slim (and also largely neglected; see Anichtchenko [2016] for a comprehensive overview of the archaeological evidence of watercraft and Hill [2023, 2024] on their cultural significance), a small body of figurative art depicts boating (see Bettina Paulsson in Chapter 4 of this volume for an analogous exploration of rock art depicting maritime harvesting). In the Eastern Arctic this takes the form, predominantly, of representations on incised tool handles of crewed umiaks (*umiat*) and kayaks (*qajat*) employed on the water to harvest bowhead whales and caribou. Although various stages of the whale hunt are illustrated, from cruising in search of whales to towing a carcass to shore (preparatory and consequent steps are rarely suggested), the climax and crisis of the whaling narrative—the harpooner poised to strike a whale and then actually harpooning a whale—predominate (Whitridge 2024). Caribou hunting is usually represented as a constellation of figures a moment before a kill, including one or more kayakers, archers, or atlatl-armed hunters on foot and swimming or standing caribou. Earlier and later moments in the hunt do not appear in the precontact Eastern Arctic imagery, although stalking and butchery are depicted on north Alaskan tools (Chan 2013; Hoffman 1897). These conventional whaling and caribou hunting setups represent significant condensations of the actual breadth of activities attendant on boat use, which encompassed everything from scavenging wood for boat frames to exchanging the surplus from maritime harvesting (both activities undoubtedly employed watercraft themselves). Following brief considerations of action ontologies and Inuit–animal–thing relations, archaeological evidence bearing on Inuit understandings of watercraft in the Eastern Arctic is surveyed. Depictions of kayaks, archers/darters, and caribou in the Inuit archaeological record are extracted from the literature, interpreted with respect to the inferred ontologies of harvesting, and similarities and differences with the pictorial whaling discourse adduced. Rather than "flat" actor-networks of equally meaningful nodes and linkages, precontact Inuit depicted, and presumably imagined, the extraordinarily complex entailments of watercraft use in terms of particularly meaningful actantial hinges—archetypically, boat-borne hunters encountering swimming prey—implying a distinct ontology premised on these repeated moments of hunter–prey encounter.

Actor-Network Theory, Ontologies, and Figurative Art

The two conceptual frameworks mentioned here—actor-network theory and ontology—have gained variable theoretical traction in archaeology. Actor-network theory imports semiotician Algirdas Greimas's account of the unfolding narrative action that underlies text to the social and material worlds, construing the arrangements and interactions among people, animals, things, and immaterial phenomena as endlessly emergent actantial structure (Greimas 1987; Latour 1993, 2005; Law and Hassard 1999). The particular composition of these arrangements—the intricate narrative symbology stitched into an Iglulingmiut angagok's parka (Boas 1907: 508–510), swarms of mask-like Dorset faces carved into living soapstone at an outcrop in arctic Quebec (Arsenault 2013)—speak to the idiosyncratic configurations of the cultural universes from which they emerged, so the actor-network is a useful way of articulating a ubiquitous and essential ontological eccentricity. Far from being opaquely inaccessible, however, archaeologists gather a wealth of contextual evidence that helps delimit and describe these historically evolving worlds, whether the social relations materialized in dwelling and settlement layout or the intricate mirroring of animal behavior in the elaborate technological setups baked into harvesting gear (Whitridge 2004).

Ontology, for its part, has become one of the theoretical metanarratives of contemporary archaeology—part of archaeology's *own* ontology. Coherent with actor-network theory, the basic premise of an archaeological ontology is that peoples in the past (as in the present) inhabited a world thoroughly configured by the ways that their precursors imagined and experienced reality; these configurations were deeply sedimented in people's memories, understandings, and bodily habitus, not to mention their material things and immaterial relations. This implies not the sort of singular, universal, realist cosmos traditionally posited by Western science but local, idiosyncratic realities inhabited by distinctive sorts of people, other living entities, physical matter, and assorted forces, any of which might be understood to be sentient and conversant with humans and other interlocutors (Alberti 2016; Viveiros de Castro 1998). For those fully inhabiting such a world, the forms and agencies that are understood to govern everyday experience may be sufficiently dissimilar to our own that archaeologists are liable to misconstrue the resultant material record, so an ontologically attentive archaeology strives to discern, articulate, and interpret these understandings subject to the limitations of its own historically idiosyncratic situatedness as a scholarly discourse on the material traces of past lifeworlds.

Luckily, archaeologists do not have to start from scratch. People in the past reflected on their world and its historical antecedents, producing not only a durable material discourse but also often tangible metadiscursive commentaries on it in the form of vibrant bodies of ornamentation and figuration. Hence, actantially sophisticated objects like kayaks, with dozens of distinctive components and accessories and correlatively complex fields of articulation with their human users and nonhuman settings, incorporate a long-running sociotechnical history that can be divined in the niceties of their design but are also represented in figurative art, where they are meaningfully situated alongside other entities like weapons and animals. Decoration, for its part, is not merely some sort of ornamental veneer on material culture but an essential semiotic constituent of decorated things, a thick layering-on of additional meanings that helps define them and guide their interpretation and use. In the present case, I imagine how precontact Inuit in the Eastern Arctic conceptualized the world *they* inhabited by way of their representations of the activities that transpired within it, and in particular the representation of watercraft employed, iconically, in the course of harvesting animals. Watercraft were foundational to Inuit economy and, along with the dog-drawn komatik (Ameen et al. 2019; Losey et al. 2018; Whitridge 2018), were a cornerstone of the mobility tactics that allowed Inuit to rapidly colonize the vast expanse of the Eastern Arctic between about 1200 and 1450 CE. By way of these assemblies Inuit aligned themselves with animals and things to compose the idiosyncratic networks that organized their lives (Hill 2023; Whitridge 2004).

Inuit–Animal–Thing Relations

Most precontact Inuit economies in the Eastern Arctic—arctic Canada east of the Mackenzie Delta as well as Greenland—were tightly focused on maritime harvesting, especially of ringed, harp, and bearded seals; walrus; and beluga and bowhead whales (Betts 2016; Savelle and McCartney 1988). These yielded much of the food, combustible blubber, hides, ivory, baleen, and bone that fed and furnished the Inuit lifeworld. Other taxa were locally important, such as fish, birds, and musk ox (Whitridge 2001), but only caribou was widely regarded as an essential economic complement to marine mammals as a source of food, sinew, antler, and, especially, hides for winter clothing (Stenton 1991). The habitability of the Eastern Arctic was utterly conditional on tailored hide clothing, including waterproof gear for operating boats during the open-water season (the hulls of which were themselves covered with stitched hides) and insulating clothes for winter (Issenman 1997). The gear was manufactured from different kinds of animal skin or gut for different purposes and subject to availability, but caribou hides harvested in late summer or early fall, after warble fly perforations had healed and before a thick winter coat had formed, were considered superior for winter garments and procured through concerted harvesting efforts and trade (Burch 1988).

Certain marine mammal species could be harvested without watercraft at certain times of year, principally the ringed seal in winter at its sea-ice breathing holes, but Inuit depended on large open-hulled umiaks and small closed-hull kayaks to harvest the rest. Perhaps counterintuitively, caribou were also preferentially harvested from kayaks as they crossed rivers, lakes, and channels on their annual migrations, typically in concert with terrestrial drive systems that helped hunters channel animals' movements toward tactically advantageous water crossings (Friesen 2013; Stewart et al. 2004). Watercraft were pivotal not only to Inuit economy but to sociality and belief, and, like the iconic implements used to process food and manufacture other tools—the woman's knife (*ulu*) and man's knife (*sapik*)—were characteristically gendered, as woman's boat (umiak) and man's boat (kayak). Elaborate bodies of symbolism traditionally surrounded the vessels and their uses (Arima 1991, 2004; Heath and Arima 2004; Petersen 1986) and even the parts of which they were composed (Hill 2024), understandings that have proved critical for identifying and interpreting archaeological remains of the watercraft themselves (Anichtchenko 2016, 2017).

Precontact Inuit Watercraft

In addition to boat parts and accessories (like paddles and boat racks), the equipment employed from them, the economic footprint of their use reflected in site distributions and harvesting refuse, and the paleopathological ramifications of cold-water boating, depictions of precontact Inuit watercraft in various media are essential for archaeologically situating these complex material cultural assemblies within local Inuit ontologies (Walls 2014: 46–56). Walls reviews the Eastern Arctic archaeological evidence for kayaking and further provides an extended exploration of the intricate complex of knowledge and technical skills that inform contemporary kayak production and use in Greenland Inuit society. Substantially complete examples of a fifteenth-century umiak frame from Peary Land (Knuth 1952; Petersen 1986) and an early historic or protohistoric kayak from Morris Bay (Walls et al. 2016), both in northern Greenland, are exceptional examples of the first, given the challenges Inuit shipbuilders faced in accumulating sufficient driftwood for a boat frame and the enormous labor that then went into manufacturing one. Substantial components of repaired or dismantled craft were then liable to be rendered unidentifiable through reduction and recycling of valuable wood and hide coverings for other purposes. Isolated or fragmentary elements of recognizable ribs, stringers, seats, cockpits, and so forth have been sporadically identified in the past (e.g., Arima 2004; Mathiassen 1927a; Walls et al. 2016), but Anichtchenko's (2016) careful reanalysis of archaeological collections suggests that there is a much more extensive material record of frame parts than previously realized. Key accessories such as deck fittings (Walls et al. 2016), paddles (particularly their bone or ivory edging; Gulløv 1997), deck scrapers and amulets (Walls et al. 2016), and stone rests for elevating and protecting both umiaks and kayaks when not in use (Savelle 1987) have also been reported from a number of sites. Open-water sea mammal hunting gear (especially parts of harpoons, darts, lances, and floats) is ubiquitous in precontact Inuit assemblages, but the successful operation of watercraft is perhaps best reflected in the consistent zooarchaeological evidence for open-water hunting. Faunal refuse associated with warm-weather sites (and seasonality determinations of cached resources) points to the reliable procurement of baleen whales and other marine mammals employing watercraft (Betts 2016; Savelle and McCartney 1988, 1994) and the organization of seasonal settlement rounds (including offshore travel) to enable this (Savelle 1987).

A distinct category of evidence for watercraft is their nonfunctional (with respect to their immediate operation) depiction, at least three important genres of which have been encountered archaeologically in the Eastern Arctic: simple wooden, ivory, or baleen models of watercraft usually interpreted as toys, stone outlines of kayaks on dry land employed in games and training, and incised depictions on tool handles of various activities, typically focused on harvesting, that often include watercraft. Each of these implies a rich discourse on watercraft targeted, respectively, at children, adolescents and young men, and adult men, although all of them would have circulated to varying degrees within mixed gender and age settings. Miniature representations of watercraft in wood, ivory, and baleen were presumably usually intended for children's play. Eastern Arctic examples of wood seem large and rugged enough for active play (11 complete toy kayaks from South West Greenland range from 61 to 200 mm and average 138 mm; Gulløv 1997: 219–221) and are generally unadorned but for an occasional inserted dowel schematically representing the kayaker (e.g.,

Holtved 1944: 284–285; Mathiassen 1927a: 46, 163; McCullough 1989: 213–214; McGhee 1984: 72; Schledermann and McCullough 2003: 102). Although aesthetically spare, they are often well crafted and therefore, like the miniature wooden figurines conventionally interpreted as dolls (Whitridge 2021), were likely manufactured by adults. Baleen examples are more schematic, taking the form of simple cut-outs that resemble watercraft in plan or profile (e.g., Holtved 1944: 284; McCullough 1989: 213); if not for the complementary baleen cut-outs of double paddles, the kayaks in plan would be difficult to recognize. Western Arctic examples of ivory kayak figurines (e.g., Anichtchenko 2017: 33–36; Arima 2004: 137–141), some with integral sculpted kayakers and inflated sealskin floats, would have been substantially more time-consuming to produce than wood or baleen ones but could conceivably also have been playthings, but with the marked disadvantage (for toy boats) of not being buoyant.

Simple cobble outlines of kayaks and umiaks much smaller than the actual versions recall the dwelling outlines that are interpreted as children's playhouses (e.g., Hardenberg 2010; Holtved 1944; Schledermann 1975: 121) and may have figured in the imaginative play of younger children. However, kayak outlines have sometimes been reported in groups where, based on Inuit accounts, they were employed as part of a skilled game that trained youths and young adults in accurately throwing a dart while seated with legs straight, as if paddling the figured kayak (Walls 2012). Both the miniature kayaks and umiaks and scaled-up outlines speak to fields of enculturation of the young to the adult activities of operating a skin vessel on the water. Like the dolls and household equipment that oriented girls' play toward the management of social relationships and the operation of an arctic domicile, boys' play was analogously equipped for harvesting and traveling (although both boys and girls sometimes played with both sorts of toys; Park 1998). Perhaps because kayak hunting was such an exceptionally skilled activity, demanding both mastery of a challenging watercraft in life-threatening conditions and the proficient operation of atlatl, dart, harpoon, and lance in hunting swimming prey, the material record of training is substantial and archaeologically recognizable.

Incised depictions represent a specialized genre of script-like hunting discourse that recognizably extends over more than a thousand years of Inuit history (Chan 2013). They often occur on manufacturing equipment, such as drill bows and knife handles, and so would have been produced and observed in the intensely gregarious arena of the *qargi* or men's house, where hunters manufactured and repaired harvesting equipment, shared stories, performed, gamed, feasted, and enacted harvesting-related and other rituals (Hoffman 1897; Ray 1982; Whitridge 2024). Stick figure depictions of scenes of everyday life, with a decided emphasis on core economic activities such as whaling, caribou hunting, and sealing, represent an occasional genre of figurative decoration of ivory and antler tools from at least early Western Thule times (ca. 950–1300 CE) in northern Alaska and from the Classic precontact Inuit period (ca. 1200–1450 CE) in the Eastern Arctic. These supplemented the elaborate geometric and curvilinear decoration that was so characteristic of earlier Old Bering Sea, Birnirk, and Punuk material culture but was becoming increasingly scarce by Western Thule times, and the occasional integration of three-dimensional representations of animals and humans into tool design. Engraved, figurative line art seems to have exploded in popularity in the Western Arctic in the early historic period, judging from the abundance of decorated drill bows and other tools in nineteenth- and early twentieth-century ethnographic collections (Bockstoce 1977; Chan 2013; Hoffman 1897). Together with the more occasional precontact specimens, these constitute an important window on Inuit, Iñupiat, and Yup'ik graphical discourse and everyday life.

Prominent within this genre are depictions of harvesting that typically include hunters, their watercraft and weapons, and prey. Although this seems like a straightforward artistic subject for hunting-dependent groups, it is important to recognize that harvesting involved much more than meeting prey on the tundra or ocean, weapon in hand. As the toys and gaming setups imply, years of observation and practice preceded an individual's transition to the role of successful hunter. All of the gear employed in the harvest had first to be produced from painstakingly assembled materials; the activities of other community members, and especially other hunters, had to be logistically coordinated; and communities typically engaged in a protracted cycle of rites, festivals, and observances that maintained their relations with animal prey, deities, and each other throughout the year. Out on the land, water, or sea ice, hunting entailed the correct reading and anticipation of conditions and prey behavior, operation of sophisticated equipment and facilities, and synchronization of the movements of hunters and prey. Once an animal was harvested, it frequently had to be transported to camp or shore and then processed before being consumed, often communally, and its various products stored or converted into critical items of material culture and perhaps traded. All of these would equally have been candidates for depiction, and, indeed, sometimes were (Chan 2013: 604–639 catalogs scores of such themes on Alaskan drill bows), but the moment of prey encounter on the water was a singularly popular subject of incised art. These contemporary representations of the lives of hunters and game are thus informative for what they leave out as much as for what they show, displaying a meaningfully myopic focus on a small subset of harvesting-related activities and so revealing the inner perceptions and motivations of the artist.

Archaeological Depictions of Kayak(er)s, Caribou, and Archers

Concrete representations of watercraft include incised depictions of harvesting scenarios on bone, antler, and ivory tool handles that occur at low frequencies on ancestral Iñupiat and Inuit sites throughout the North American Arctic after 1200 CE and very occasionally on earlier sites in the Western Arctic. This modest prehistoric sample is vastly enlarged from the mid-nineteenth century, when collectors began to acquire a wide range of northern material culture for museums in the United States, Canada, and Europe and a lively tourist market emerged for decorated ivory tools and figurative art (Bockstoce 1977; Chan 2013; Hoffman 1897; Ray 1982). The content of this imagery is archaeologically informative from a number of perspectives. Some tableaux consist of item-ized tallies of hunted animals signified by skins or body parts, thus speaking to the interests and values of their makers as well as a maturing pictographic shorthand. Even more informatively, many seem to depict everyday events in the lives of their creators, such as scenes of village life and realistic hunting scenarios. Chan (2013) assembled an enormous sample of such figurative engravings from coastal northwest Alaska, mostly dating to the late nineteenth and early twentieth century, which provides a valuable window on the art's production context and content (as well as illustrating a substantial portion of it). Bowhead whaling and walrus hunting are particularly well represented, but all sorts of harvesting scenarios are depicted, including caribou hunting. This is a remarkable dataset in that, unlike the extensive bodies of ethnographic and photographic documentation of Yup'ik, Iñupiat, and Inuit on the eve of their absorption into the global network, it articulates their lives and interests as they perceived them, autoethnographically, in a familiar and long-standing graphical idiom (although Inuit also occasionally engaged in both ethnography and docu-mentary photography, as exemplified by the important early twentieth-century work of the ethnographer Knud Rasmussen and photographer Peter Pitseolak). Approaching Inuit–animal–thing relations from the perspective of their local representation allows Inuit figurative art to guide the archaeological narrative.

The Alaskan record seems critical for approaching the Classic precontact and historic material from the Eastern Arctic since it was groups from north-ern Alaska that expanded into the Mackenzie Delta–Amundsen Gulf region at some point before 1200 CE, and then into the Eastern Arctic. The early Alaskan specimens are thus directly related to the Classic precontact examples as im-mediate historical precursors within the same cultural lineage. Chan reproduces two of the earliest examples of figurative engraving from northern Alaska, one on an ivory wrist guard collected by Vilhjalmur Stefansson at Birnirk (Chan 2013: 231–232) and the other an adze handle excavated at Kurigitavik by Henry

Collins (Chan 2013: 239–241; Table 10.1; Figure 10.1, designs not reproduced in the figure were either illegible or not illustrated in the original publication). Both are broadly dated to late Birnirk or early Western Thule times and both exhibit scenes with a single archer confronting a single caribou (Figure 10.1:1, 3; although watercraft figure in the elaborate whaling and bear-hunting scenarios on a Birnirk or earlier drill bow from Chukotka [Gusev 2022], neither caribou nor kayaks are present). Giddings and Anderson (1986: 85–86) illustrate the design on an early Western Thule bodkin from Cape Krusenstern that is slightly more complex than the earliest Alaskan tableaux and appears to depict two hunters armed with atlatls who have darted a caribou (Figure 10.1:2; no scale or measurement in original). Giddings and Anderson suggest a watercraft in the background is an umiak, but it equally resembles depictions of kayaks, especially given the size of the craft relative to the people and animal, the suggestion of a single paddler, and the depth of the prow suggested by hatching (as opposed to the projecting "horns" [Petersen 1986: 121] on an umiak usually represented by paired lines in the Alaskan incised art). Caribou hunting was clearly an important theme of incipient pictographic tool decoration, and watercraft came to play an important role in the former activity. While the earliest Alaskan examples are sparse and schematic, the one from Cape Krusenstern begins to evoke the size and complexity of harvesting setups as later toolmakers envisioned them.

Twenty archaeological instances of figurative line art depicting kayakers, caribou, or archers were identified in the Canadian Arctic record, ranging in age from early Classic precontact to early historic Inuit (Table 10.1). Five such scenes occur in distinct panels on a drill bow from northwest Baffin Island, two occur on opposite sides of a knife handle from southeast Somerset Island, two on opposite sides of a comb from northwest Foxe Basin, and two on opposite sides of an ivory fragment from northern Southampton Island. The rest occur individually. The fragmentary kayak images from Southampton Island are on a small portion of an object of uncertain size, the comb from Naujan is missing about one-third of the decorated panel, and portions of one of the caribou hunting panels on the Arctic Bay drill bow have been destroyed; the rest are complete, although Mathiassen's (1927a) photographs are grainy and sometimes difficult to interpret (hence, the published comb image from Kuk is largely illegible). The westernmost Eastern Arctic example is an ulu handle from Booth Island with decoration that appears to be unfinished (Figure 10.1:4); the figure identified as a caribou occurs only in partial outline, and the impaled animal may be a seal or other marine mammal (Morrison 1990: 92). Booth Island harpoon head types and attributes seem to span the entire Classic precontact period but include some traits (nippled shoulders, lashing

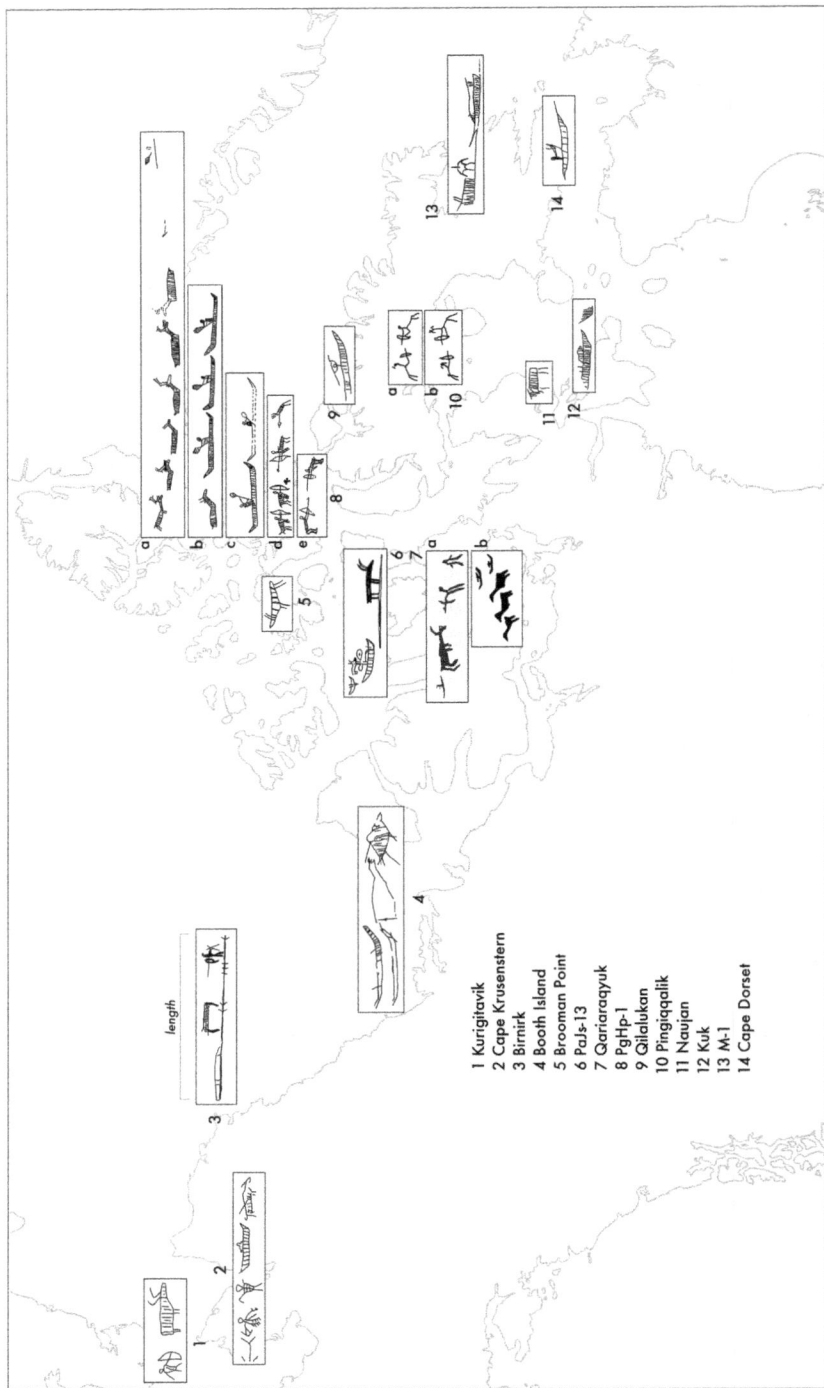

Figure 10.1. Distribution of premigration, precontact, and historic Inuit depictions of kayak(er)s, caribou, and archers.

1 Kurigitavik
2 Cape Krusenstern
3 Birnirk
4 Booth Island
5 Brooman Point
6 PaJs-13
7 Qariaraqyuk
8 PgHp-1
9 Qilalukan
10 Pingiqqalik
11 Naujan
12 Kuk
13 M-1
14 Cape Dorset

slots, vestigial spurs) consistent with the earlier part of this range. PaJs-13 and Qariaraqyuk, on southeastern Somerset Island, are contemporaneous winter villages only 10 km apart that likewise span the Classic period. The side-slotted knife handle from PaJs-13 (Figure 10.1:6) seems to depict two kayaker–caribou pairs—a larger and more detailed one in the foreground and a smaller one in the background (Habu and Savelle 1994). The end-slotted knife handle from Qariaraqyuk (Figure 10.1:7) includes a panel on one side depicting a kayaker in the middle distance and a caribou, archer, and *inuksuk* in the foreground (Whitridge 2013). On the other side are two kayakers and three taxonomically ambiguous animals; only two posterior limbs are clearly indicated for each, and so the depictions seem consistent with a seventeenth-century engraving of a Greenland Inuit kayaker hunting waterfowl with atlatl and bird spear (Settle 1675). The zoomorphic design on an ivory comb from the contemporaneous Brooman Point site (Figure 10.1:5) is also somewhat difficult to speciate but appears to depict a lone gamboling caribou (McGhee 1984, and see below).

An ivory drill bow from PgHp-1 near Arctic Bay (Figure 10.1:8) provides the most complex example of pictographic art from the precontact Canadian Arctic (Maxwell 1983). Alongside panels depicting bowhead whaling from umiaks and kayaks with photographic verisimilitude (Whitridge 2024: Figure 3.4) are scenes of village life, feuding and warfare, and two distinct panels illustrating kayakers pursuing caribou. One of these, showing three kayakers approaching a swimming caribou, is intact and the other, which appears to depict two kayakers pursuing six caribou, is partially exfoliated. Two kayakers appear in a partially eroded scene depicting bowhead whaling, and one panel on each side of the drill bow depicts archers facing each other, either singly ("feuding") or in pairs ("warring"). Of the seven caribou, two appear to have distinctly smaller antlers (very similar to the Brooman Point comb), which may reflect the out-of-phase seasonal development and different realized sizes of male and female caribou antlers; males develop earlier and significantly larger racks (Miller et al. 2023), so a pair of delicate-looking prongs appears to have stood for developing, or simply smaller, female antlers. Not only are the individual panels on this drill bow pictorially and narratively complex (i.e., the figures well delineated and precisely equipped), but their juxtaposition on the same object seems to compose a historical mosaic: the decorative border surrounding the warfare episode is discontinuous with the adjoining scenes depicting caribou hunting (along one edge) and umiaks (along the other), which both lack this framing, suggesting that the different decoration zones were added in separate carving episodes. This record of the owner's (or owners') personal and community history may have accumulated over time. The decoration on a

Table 10.1. Incised Tool Decoration Depicting Kayak(er)s, Caribou, and Archers from Precontact and Early Historic Inuit Sites

Site	Map Key	Cultural Affiliation	Content	Length (mm)	Object	Source
Kurigitavik, NW Alaska	1	Birnirk/early Western Thule	archer, caribou	51	adze handle	Chan 2013: 241
Cape Krusenstern, NW Alaska	2	early Western Thule	dart, 2 hunters with atlatls, kayaker (?), darted caribou	—	bodkin	Giddings and Anderson 1986: 85–86
Birnirk, N Alaska	3	Birnirk/early Western Thule	unidentified object, caribou, archer	62	wrist guard	Chan 2013: 231–232
Booth Island, S Amundsen Gulf	4	Classic precontact Inuit	2 kayaks, caribou, impaled animal	73	ulu handle	Morrison 1991: 92
Brooman Point, E Bathurst Island	5	Classic precontact Inuit	caribou	18	comb	McGhee 1984: 74
PaJs-13, SE Somerset Island	6	Classic precontact Inuit	2 kayakers, 2 caribou	90	men's knife	Habu and Savelle 1994: 3
Qariaraqyuk, SE Somerset Island	7a	Classic precontact Inuit	kayaker, caribou, archer, inuksuk	65	men's knife	Whitridge 2013: 234
Qariaraqyuk, SE Somerset Island	7b	Classic precontact Inuit	3 waterfowl (?), 2 kayakers	42	men's knife	Whitridge 2013: 234
PgHp-1, NW Baffin Island	8a	Classic precontact Inuit	6 caribou, 2 kayakers (partially exfoliated)	200	drill bow	Maxwell 1983: 83
PgHp-1, NW Baffin Island	8b	Classic precontact Inuit	caribou, 3 kayakers	105	drill bow	Maxwell 1983: 84
PgHp-1, NW Baffin Island	8c	Classic precontact Inuit	2 kayakers	60	drill bow	Maxwell 1983: 84

Site	Map Key	Cultural Affiliation	Content	Length (mm)	Object	Source
PgHp-1, NW Baffin Island	8d	Classic precontact Inuit	4 warring archers	77	drill bow	Maxwell 1983: 83
PgHp-1, NW Baffin Island	8e	Classic precontact Inuit	2 feuding archers	38	drill bow	Maxwell 1983: 84
PgHp-1, NW Baffin Island	8	Classic precontact Inuit	caribou, kayaker, bird, person (not illustrated)	—	"dagger"	Maxwell 1983: 79
Qilalukan, N Baffin Island	9	Classic precontact Inuit	kayaker	33	comb	Mathiassen 1927a: 185
Pingiqqalik, NW Foxe Basin	10a	late precontact Inuit	2 feuding archers	30	comb	Desjardins 2017: 112
Pingiqqalik, NW Foxe Basin	10b	late precontact Inuit	2 feuding archers	31	comb	Desjardins 2017: 112
Naujan, NW Hudson Bay	11	Classic precontact Inuit	caribou (mostly complete)	10	comb	Mathiassen 1927a: 70
Naujan, NW Hudson Bay	11	Classic precontact Inuit	caribou (not illustrated)	—	comb	Mathiassen 1927a: 70
Kuk, N Southampton Island	12	Classic precontact Inuit	kayak, umiak (fragmentary)	44	unidentified	Mathiassen 1927a: 251–252
Kuk, N Southampton Island	12	Classic precontact Inuit	kayak, whale (fragmentary, not illustrated)	—	unidentified	Mathiassen 1927a: 251–252
Kuk, N Southampton Island	12	Classic precontact Inuit	2 caribou (not illustrated)	—	comb	Mathiassen 1927a: 260
M-1, SE Baffin Island	13	early historic Inuit	tent, person, kayak	44	drill bow	Schledermann 1975: 122
Cape Dorset, S Baffin Island	14	precontact/early historic Inuit	kayaker	17	unidentified	Arima 1994: 194

dagger-like ivory object from the same site is narratively likewise idiosyncratic (a bird sitting on the stern of a kayak while the kayaker approaches a caribou, and a figure on shore with upraised arms).

An ivory comb from Pingiqqalik in northwest Foxe Basin (Figure 10.1:10) has virtually identical scenes of archers facing each other—presumably feuding—on each side (Desjardins 2017: 113). This is the only object in the sample with no depictions of watercraft or animals, but the feuding scenes are very similar to the one on the Arctic Bay drill bow. Although sharpened and barbed arrowheads (as distinct from bird blunts) might have been used against a variety of terrestrial species, other weapons and techniques were favored for polar bear, musk ox, and small game; the principal uses of sharpened or barbed arrowheads seem to have been for caribou hunting and interpersonal conflict (as clearly depicted on the specimens from Qariaraqyuk and PgHb-1, respectively). In his original monograph defining "Thule" (precontact Inuit) culture, Mathiassen (1927a) reported kayak and caribou pictographs from northern Baffin Island (Qilalukan) and northwestern Hudson Bay (Naujan and Kuk). He illustrates a single, slightly damaged caribou on one side of a comb from Naujan (Figure 10.1:11; Mathiassen 1927a: 70) but does not provide an illustration of a comparable figure reported on the reverse. Two caribou are indicated on one side of a poorly reproduced comb from Kuk (Mathiassen 1927a: 260). A comb from Qilalukan (Figure 10.1:9) illustrates a kayaker with upraised weapon on one side and a crewed umiak with standing harpooner on the other (Mathiassen 1927a: 185). A decorated ivory fragment from Kuk (Figure 10.1:12) appears to depict portions of a kayak and an umiak on one side, and a kayak and a whale on the other (Mathiassen 1927a: 251–252). Conceivably, the latter are snippets of complex harvesting setups like the ones on the Arctic Bay drill bow. Arima (1994: 194) illustrates an object from the Cape Dorset area (Figure 10.1:14) with a lone kayaker and Schledermann (1975: 122) portions of an eroded drill bow from B-1 on Cumberland Sound (Figure 10.1:13) that include a scene of an individual standing between a tent and a beached kayak. Although additional examples undoubtedly exist, this seems to represent a reasonable sampling of known depictions of kayaks, archers, and caribou from precontact and early historic Canadian Inuit sites and premigration sites in Alaska, suggesting interesting overlaps and divergences in the way these figures are represented.

In addition to the social and semantic implications of these depictions, explored in the following section, it is noteworthy that some of them appear to illustrate meaningful technical details of ancient watercraft. For comparative purposes, individual kayaks and kayakers are extracted from the panels in Figure 10.1, rescaled to the same length (or estimated length for fragmentary images), and enlarged in Figure 10.2. Some of the images are mirrored so that

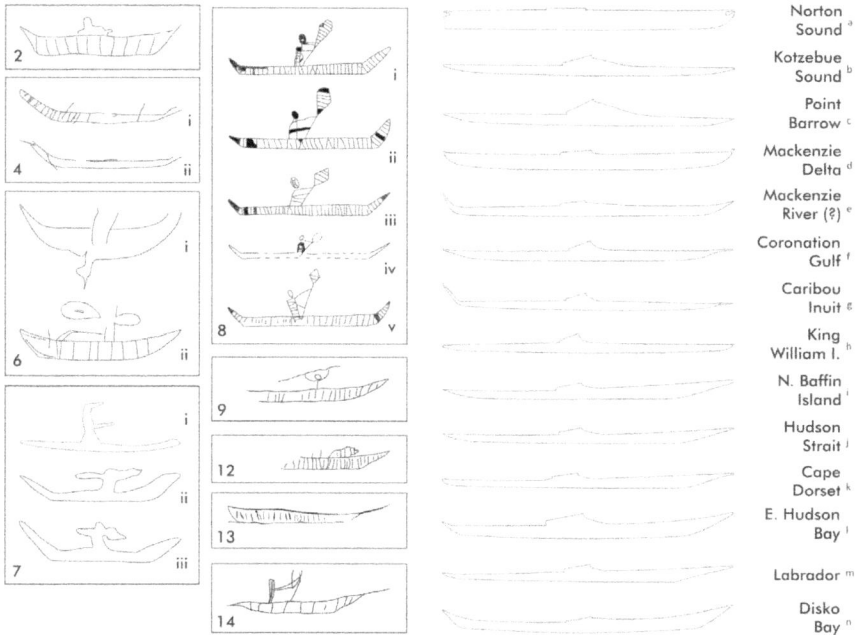

Figure 10.2. Details of archaeological kayak depictions. The images in the left two columns are rescaled to the same length (complete length estimated for 9 and 12), mirrored as necessary so that the inferred stern is at left, and numbered as in Figure 10.1. The late nineteenth- and early twentieth-century kayak profiles at right are likewise rescaled and reoriented. Notes: *a.* Chapelle 1964: 198; *b.* Chapelle 1964: 201; *c.* Chapelle 1964: 201; *d.* Chapelle 1964: 201; *e.* Chapelle 1964: 203; *f.* Chapelle 1964: 203; *g.* Chapelle 1964: 203; *h.* Chapelle 1964: 203; *i.* Arima 1990: 187; *j.* Arima 1990: 187; *k.* Chapelle 1964: 205; *l.* Arima 1990: 187; *m.* Arima 1990: 187.

all of the kayaks are depicted conventionally, with what appears to be the stern at left, and a representative array of historic kayak profiles from various regions is shown at right, likewise rescaled and aligned. A late nineteenth-century Central Yup'ik profile from the southern shore of Norton Sound, with its distinctly truncated stern and bifurcated bow, is provided by way of contrast. All of the rest are historic examples from Iñupiat and Inuit areas east of Bering Strait and resemble the archaeological depictions more closely. Although some of the latter are overly stylized (e.g., 7.i), and the orientations of others are debatable (especially 4.i and 4.ii, which are taken to resemble the Mackenzie River profile most closely but would resemble the East Hudson Bay profile if reversed), most of the illustrations provide some robust, interpretable structural detail, especially with respect to the height and inclination of the bow and stern. In particular, the inferred sterns tend to be angled sharply upward, between about 45 and 60°, from the water surface (e.g., 2, 6.i, 6.ii, 7.ii, 7.iii, 8.i, 8.iii, 8.iv, 8.v, 13), while the inferred bow tends to project further from the submerged portion of the hull and at a more acute angle, between about 30 and 45° (e.g., 2, 6.ii, 8.i, 8.iii, 8.iv, 9, 12, 13, 14). The closest analogs for most among the ethnohistoric profiles are those from the Mackenzie River (unfortunately, with somewhat uncertain provenience) and Disko Bay, although the archaeological examples from Qilalukan, M-1, and Cape Dorset more strongly resemble the ethnohistoric examples from the same regions. Kayak designs clearly varied significantly in time and space, reflecting diverse paddling conditions (wind, waves, ice) and kayak uses (caribou hunting, sealing, whaling) as well as regional cultural traditions and shifting styles. Although it would be challenging to characterize these from the archaeological record of boat parts alone, a fine-grained analysis of boat forms can usefully integrate these two-dimensional depictions, along with the three-dimensional toys, paddles, and boat rests that Arima (2004) effectively explored.

Harvesting Ontologies and the Semiotics of Tool Decoration

The motifs of interest—kayaks, kayakers, archers, darters, caribou—are depicted in these panels both singly and in various combinations. Kayakers occur apart from hunters and caribou (sometimes with whales, waterfowl, or village scenes) in seven instances, caribou alone in four, and archers alone in four. Kayakers and caribou occur together five times, and archers and caribou twice, and all three figures occur together twice. Figure 10.3 illustrates these proportions schematically. Caribou, both alone and in combination with the other motifs, and kayaks were popular subjects of figurative tool decoration; caribou occur in 54% of the representations that include one of these three motifs, and kayakers in 58%. Perhaps because kayaks were employed in all facets of warm-weather travel and harvesting, including whaling and fowling, they were likely to figure in the scenarios that carvers were inclined to depict. Kayakers were critical to caribou hunting in many areas, deploying in lakes, rivers, and inlets to intercept animals funneled to crossings by *inuksuit* and beaters (Friesen 2013).

Kayaks are unusual technical devices (Arima 1975, 1987; Heath and Arima 2004; Petersen 1986; Walls 2014). Not only are they exceptionally effective and useful arctic watercraft (as, indeed, are umiaks), but operating one requires the kayaker to slide their legs deep inside the hull, allowing them to effectively transfer muscular energy to it but rendering their legs otherwise immobile; in a practical, physiological fashion, the kayak substitutes for the user's legs, creating a new person-kayak chimera that can maneuver smoothly and rapidly on the water's surface. This effect is accentuated when the kayaker's waterproof clothing (i.e., a tightly hooded gutskin parka) is affixed to the cockpit, allowing a skilled paddler to safely roll the craft (Heath 2004; Walls 2014). This coalescence of kayaker-person and kayak-thing is exemplified ethnographically by Greenland Inuit hunters who identified so closely with their role as kayakers that they would sometimes arrange to have themselves interred seated in their craft, as if kayaking into the afterlife (Whitridge and Kleist 2024). However, transforming into a person-kayak was not without its risks; a variety of disturbing psychological effects, especially losing the perception of the horizon under still, overcast conditions, could induce panic disorder and led some hunters in the past to renounce kayaking altogether (Christensen and Rud 2013; Heath 2004). The frequent depiction of kayaking in Inuit incised art, along with widespread stone outlines and toys (and occasional person-kayak carvings), speak to the substantial set of understandings, skills, and anxieties that emerged over a lifetime of familiarity with the craft as well as the cultural discourses that accompanied this.

The most common decorated tool types in this sample are drill bows, men's knives, and combs, with single examples of the other types (Table 10.1). Although scenes of feuding occur on both sides of one comb, and kayak and umiak hunting on opposite sides of another, the most common thematic element in the comb art is caribou, either paired or alone. Combs are conventionally understood as objects used by women in the sort of elaborate hair dressing documented historically (e.g., Boas 1964 [1888]: 150–151) and suggested by prehistoric figurative representations (including dolls and anthropomorphic varieties of combs and *tingmiujat* [flat-bottomed game pieces resembling a floating bird or chimerical person-bird]). While caribou hunting and various scenes involving kayaks occur on an array of tool types mainly associated with men's work, the association of isolated caribou with women's personal dress is striking. One practical connection that can be drawn here is between women's lifeworld and the land (as opposed to their oft-cited symbolic association with the sea; McGhee 1977). While men regularly pursued swimming prey (including caribou) from watercraft during the open-water season, women's routines at this time were closely bound up with activities on shore, especially processing the flesh, fat, hides, and bone of harvested animals. While caribou, from this perspective, may have suggested the drudgery of scraping hides and pounding bones for grease, when viewed at a distance or contemplated in retrospect, they may have held a positive affective resonance, evoking the warmth and companionship of the summer camp, the freedom of roaming the tundra in search of berries and herbs, or other pleasant associations. This is suggested by the ornamental caribou's detachment from any work context (such as being pursued by archers and kayakers, as in the decoration of men's tools) in the comb's isolated visual field. Furthermore, if men made and decorated the combs, as seems possible, then pairing this sort of imagery with an intimate personal object that may have been used by women speaks also to the emotional valence of women's and men's relations, in a similar way to women's production of men's clothing.

The depiction of activities ethnographically (and conventionally) associated with men (feuding and maritime harvesting) on two of the combs is more surprising and is challenging to interpret. It may represent women's interest or actual engagement in these activities or, alternatively, the combs' possession and use by men, although the ethnographic literature often refers only to women using combs (e.g., Boas 1964 [1888]: 150–151; Jenness 1946: 50–51). Rasmussen (1931: 264), however, mentions Netsilingmiut taboos preventing "the people of a village" from combing their hair while in mourning, apparently implying that men normally did as well. Oosten (1982: 104) further draws attention to the centrality of combing in Inuit Sedna traditions: when game disappears, an angakok (who would usually be male) travels to the sea goddess's undersea home and combs her hair to appease her. Eastern Arctic combs are often exceptionally well made in relatively precious ivory and are finely decorated with patterned borders and various geometric and figurative motifs, but archaeological discussion of their use and meaning is surprisingly thin (e.g., Mathiassen 1927b: 113–115). They seem to have held a distinctive cultural position, part of which, given the occurrence of motifs relating to both women and men (isolated caribou versus feuding and maritime hunting), was perhaps the bridging of their respective symbolic fields. The overall abundance of scenes of archers in conflict is also intriguing but again hard to interpret. Although this only occurs on two objects in the sample, it is repeated on both, suggesting some special cultural salience to the image and to the activity. A scene of archers or darters confronting caribou likewise occurs in four panels but on four different objects; hence, it was an even more widely circulating motif. As suggested by the Cape Krusenstern and Qariaraqyuk hunting scenes, hunters (both archers and darters) were also linked to kayakers through their joint participation in the caribou hunt, suggesting the three composed a small but meaningful semantic nexus (Figure 10.3).

A comparison of these designs with precontact and early historic incised line art illustrating various facets of the bowhead whaling operation is also instructive. Among a sample of 139 such depictions from precontact and historic Inuit, Iñupiat, and Yu'pik contexts, a limited number of conventional scenarios were repeatedly evoked, and of these the scene most commonly represented

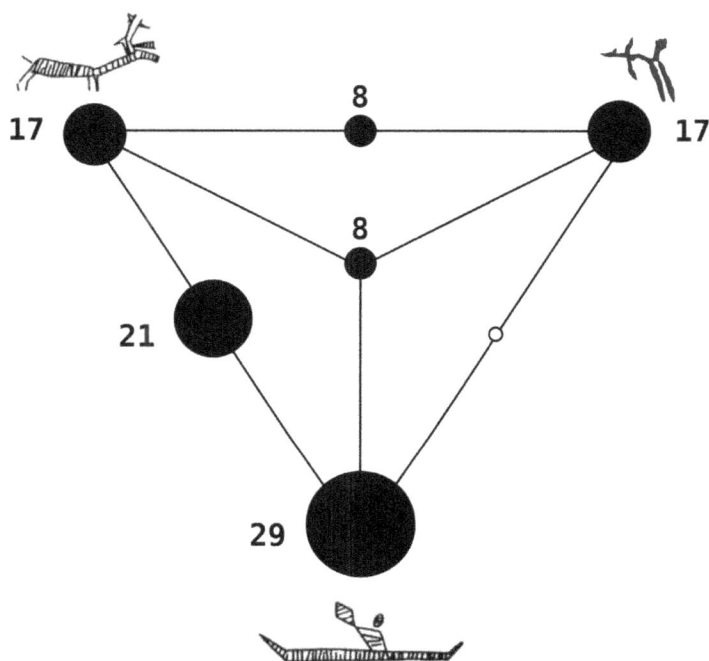

Figure 10.3. Relative frequency (%) of kayak(er), caribou, and archer motifs, alone and in combination.

was the precise instant at which a harpooner, standing in the bow of an umiak, actually struck a whale (Whitridge 2024). Of all the hours that must have been spent cruising leads, approaching potential targets, attempting to harpoon them, chasing a whale to which floats had finally been attached, killing it with lances, and then towing the carcass to shore, the fleeting moment during which the harpooner plunged the weapon into the whale's back was returned to again and again. The frequency of depiction of principal episodes in the whaling sequence reveals a steady increase in scenes approaching the climax (the standing harpooner about to strike) and then peaking at the narratological crisis (the harpoon held in contact with the whale), followed by a sharp decline in representations of the apparently anticlimactic activities of killing and towing the harpooned animal (Whitridge 2024: Figure 3.8). Although the present sample is much smaller, there is a similar tendency toward a modal, archetypal setup in the caribou hunting depictions in the form of one or a few kayakers and swimming or standing caribou, occasionally with a prominent archer or darter in the scene. Weaponry is somewhat less common than kayaks: 19 kayaks appear in 12 different scenes, whereas 15 archers/darters appear in 9 different scenes. However, archers/darters appear alongside caribou in only 4 of the latter. Unlike the whaling case, the crisis of animal struck by weapon is an infrequent theme in the caribou imagery (1 of 20 animals in 13 scenes), but in both genres hunters are represented at the apex of technical preparedness, on the prow of a crewed umiak, harpoon in hand, or chimerically merged with a kayak hull, paddle upraised. For the skilled and well-armed kayaker, perhaps this moment of close approach was the point at which a kill was effectively ensured, as it was for whalers only when the harpoon line was actually affixed to the prey.

There is a simplicity, a semiotic economy, to these images and scenarios, meant to fit in the materially confined space of a tool handle and evoking a similarly compact emotional register: the sudden copresence of human and animal on the cusp of prey being dispatched. Standing behind them, undoubtedly, are real encounters with real animals, which are recalled and distilled into the decoration of an everyday tool, and kayaks or umiaks are essential to many of these scenarios. Watercraft were pivotal, helping to compose the actantial hinge of core cultural narratives. They were both vessels for the narrative action and embodied its consequences since they were manufactured from animal skins that were either harvested directly or obtained in trade with the fruits of the harvest.

Conclusion

The archaeological record of Inuit kayak use is fairly substantial, ranging from fragments of the watercraft themselves (Anichtchenko 2016, 2017; Arima 2004; Walls 2014; Walls et al. 2016) to musculoskeletal pathologies arising from their use (Merbs 1983). Each category of evidence represents its own intriguing rabbit hole, and many (like toys) have received only cursory attention. The depiction of kayaking, here considered in its overlap with caribou hunting, is one of these neglected categories. Nevertheless, Arima (2004), Maxwell (1983), and others have demonstrated that there are layers of granular, interpretable content to the archaeological imagery, as the present study sought to explore. As with the analogous whaling imagery (Whitridge 2024), the incised kayaking line art forces us to center our understanding of the Inuit imaginary on a specific chronotope: the moment of encounter between hunter and prey. Of all the conceivable circumstances that carvers might have chosen to depict, this was the moment they returned to again and again, as storytellers and performers undoubtedly did as well. A fragment of precontact Inuit ontology might be extracted from this actantial kernel, emphasizing the overarching significance of the momentary copresence of hunter and game and all that this implies, including the importance of a deferential respect performed into reality by a lifetime of attending to and honoring game; the necessity of amassing a wealth of experiential knowledge of animal behavior, weather, and equipment; and the vicissitudes, nevertheless, of animal behavior, weather, and equipment. Interestingly, what we do not see in the engravings are suggestions of all the things that might go wrong, of arrows and darts that missed their mark or broke, of boats overturned in the heat of pursuit. The representations evoke an ontology of positive affect, of things that had gone right to enable this felicitous culmination of learning, experience, and work. Neither do these images represent anything close to a complete account of human–animal interaction since they fail to document many of the activities that came before and after. Numerous individuals besides paddlers, harpooners, and archers assumed tactically essential roles—especially the women who manufactured much of the gear and processed the spoils—but are elided in the visual narratives. The outlines of precontact Inuit ontology emerge from these highly selective depictions of the harvest, with the encounter between hunter and prey, mediated by watercraft and weapons, pointedly distilling a messy, sprawling reality.

Acknowledgments

Many thanks to Alberto García-Piquer, Mikael Fauvelle, and Colin Grier for inviting me to contribute to this volume. I am glad to be a part of a volume that honors Ken Ames, who made such a profound contribution to the social archaeology of complex hunter–fisher–gatherers.

References Cited

Alberti, Benjamin. 2016. Archaeologies of Ontology. *Annual Review of Anthropology* 45: 163–179.

Ameen, Carly, Tatiana R. Feuerborn, Sarah K. Brown, Anna Linderholm, Ardern Hulme-Beaman, Ophélie Lebrasseur, Mikkel-Holger S. Sinding, et al. 2019. Specialized Sledge Dogs Accompanied Inuit Dispersal Across the North American Arctic. *Proceedings of the Royal Society B* 286(1916): 20191929.

Anichtchenko, Evguenia. 2016. Open Passage: Ethno-Archaeology of Skin Boats and Indigenous Maritime Mobility of North American Arctic. PhD dissertation, Centre for Maritime Archaeology, Faculty of Humanities, University of Southampton, Southampton.

Anichtchenko, Evguenia. 2017. Reconstructing the St. Lawrence Island Kayak: From Forgotten Watercraft to a Bering Sea Maritime Network. *Alaska Journal of Anthropology* 15(1–2): 5–48.

Arima, Eugene. 1975. *A Contextual Study of the Caribou Eskimo Kayak.* National Museums of Canada, Ottawa.

Arima, Eugene. 1987. *Inuit Kayaks in Canada: A Review of Historical Records and Construction.* National Museums of Canada, Ottawa.

Arima, Eugene. 1990. East Canadian Arctic Kayak. *Arctic* 43(2): 187–189.

Arima, Eugene. 1994. Caribou and Iglulik Inuit Kayaks. *Arctic* 47(2): 193–195.

Arima, Eugene. 2004. Kayaks of the East Canadian Arctic. In *Eastern Arctic Kayaks: History, Design, Technique,* edited by John Heath and Eugene Arima, pp. 111–148. University of Alaska Press, Fairbanks.

Arima, Eugene (editor). 1991. *Contributions to Kayak Studies.* University of Ottawa Press, Ottawa.

Arsenault, Daniel. 2013. The Aesthetic Power of Ancient Dorset Images at Qajartalik, a Unique Petroglyph Site in the Canadian Arctic. *Boletín del Museo Chileno de Arte Precolombino* 18(2): 19–32.

Betts, Matthew. 2016. Zooarchaeology and the Reconstruction of Ancient Human-Animal Relationships in the Arctic. In *The Oxford Handbook of the Prehistoric Arctic,* edited by T. Max Friesen and Owen Mason, pp. 81–108. Oxford University Press, Oxford.

Boas, Franz. 1907. *Second Report on the Eskimo of Baffin Land and Hudson Bay.* Bulletin of the American Museum of Natural History, Vol. 15, Part 2. American Museum of Natural History, New York.

Boas, Franz. 1964 [1888]. *The Central Eskimo.* University of Nebraska Press, Lincoln.

Bockstoce, John. 1977. *Eskimos of Northwest Alaska in the Early Nineteenth Century.* Pitt Rivers Museum Monograph Series No. 1. Pitt Rivers Museum, Oxford.

Burch, Ernest S., Jr. 1988. Modes of Exchange in North-West Alaska. In *Hunters and Gatherers 2: Property, Power, and Ideology,* edited by Tim Ingold, David Riches, and James Woodburn, pp. 95–109. Berg, Oxford.

Chan, Amy. 2013. Quliaqtuavut Tuugaatigun (Our Stories in Ivory): Reconnecting Arctic Narratives with Engraved Drill Bows. PhD dissertation, Arizona State University, Tempe.

Chapelle, Howard I. 1964. Arctic Skin Boats. In *The Bark Canoes and Skin Boats of North America,* edited by Edwin T. Adney and Howard I. Chapelle, pp. 174–211. Smithsonian Institution, Washington, D.C.

Christensen, Ivan, and Søren Rud. 2013. Arctic Neurasthenia: The Case of Greenlandic Kayak Fear 1864–1940. *Social History of Medicine* 26(3): 489–509.

Desjardins, Sean. 2017. A Change of Subject: Perspectivism and Multinaturalism in Inuit Depictions of Interspecies Transformation. *Études Inuit Studies* 41(1–2): 101–124.

Friesen, T. Max. 2013. The Impact of Weapon Technology on Caribou Drive System Variability in the Prehistoric Canadian Arctic. *Quaternary International* 297: 13–23.

Friesen, T. Max, and Owen Mason (editors). 2016. *The Oxford Handbook of the Prehistoric Arctic.* Oxford University Press, Oxford.

Giddings, J. Louis, and Douglas D. Anderson. 1986. *Beach Ridge Archeology of Cape Krusenstern.* Publications in Archaeology No. 20. National Park Service, U.S. Department of the Interior, Washington, D.C.

Greimas, Algirdas. 1987. *On Meaning: Selected Writings in Semiotic Theory.* Frances Pinter, London.

Gulløv, Hans Christian. 1997. *From Middle Ages to Colonial Times: Archaeological and Ethnohistorical Studies of the Thule Culture in South West Greenland 1300–1800 AD.* Dansk Polar Center, Copenhagen.

Gusev, Sergey V. 2022. The Onset of Maritime Adaptations in Eastern Chukotka and the Emergence of Marine Economies and Seafaring Activities Between 8000 and 3500 Years Before Present. In *Maritime Prehistory of Northeast Asia,* edited by Jim Cassidy, Irina Ponkratova, and Ben Fitzhugh, pp. 291–313. Springer, Singapore.

Habu, Junko, and James M. Savelle. 1994. Construction, Use, and Abandonment of a Thule Whale Bone House, Somerset Island, Arctic Canada. *Daiyonki Kenkyu [Quaternary Research]* 33(1): 1–18.

Hardenberg, Mariane. 2010. In Search of Thule Children: Construction of Playing Houses as a Means of Socializing Children. *Danish Journal of Geography* 110(2): 201–214.

Heath, John. 2004. Kayaks of Greenland. In *Eastern Arctic Kayaks: History, Design, Technique,* edited by John Heath and Eugene Arima, pp. 5–44. University of Alaska Press, Fairbanks.

Heath, John, and Eugene Arima (editors). 2004. *Eastern Arctic Kayaks: History, Design, Technique.* University of Alaska Press, Fairbanks.

Hill, Erica. 2023. Watercraft as Assemblage in the Western Arctic. In *Sacred Nature: Animism and Materiality in Ancient Religions,* edited by Nicola Laneri and Anna Perdibon, pp. 17–32. Oxbow, Oxford.

Hill, Erica. 2024. Whales, Whaling, and Relational Networks in the Western Arctic. In *Reimagining Human–Animal Relations in the Circumpolar North,* edited by Peter Whitridge and Erica Hill, pp. 88–112. Routledge, New York.

Hoffman, Walter J. 1897. *The Graphic Art of the Eskimos, Based upon the Collections in the National Museum.* Report of the United States National Museum for 1895: 739–968. Government Printing Office, Washington, D.C.

Holtved, Erik. 1944. *Archaeological Investigations in the Thule District.* Meddelelser om Grønland, Bd. 141, No. 1–2. Copenhagen.

Issenman, Betty Kobayashi. 1997. *Sinews of Survival: The Living Legacy of Inuit Clothing.* University of British Columbia Press, Vancouver.

Jenness, Diamond. 1946. *Material Culture of the Copper Eskimo.* Report of the Canadian Arctic Expedition 1913–18, Vol. 16. Edmond Cloutier, King's Printer and Controller of Stationery, Ottawa.

Knuth, Eigil. 1952. An Outline of the Archaeology of Peary Land. *Arctic* 5(1): 17–33.

Latour, Bruno. 1993. *We Have Never Been Modern.* Harvard University Press, Cambridge.

Latour, Bruno. 2005. *Reassembling the Social: An Introduction to Actor-Network Theory.* Oxford University Press, Oxford.

Law, John, and John Hassard (editors). 1999. *Actor Network Theory and After.* Blackwell, Oxford.

Losey, Robert, Robert Wishart, and Jan Peter Laurens Loovers (editors). 2018. *Dogs in the North Stories of Cooperation and Co-Domestication.* Routledge, New York.

Mathiassen, Therkel. 1927a. *Archaeology of the Central Eskimo, I: Descriptive Part.* Report of the Fifth Thule Expedition 1921–24, Vol. 4, No. 1. Gyldendalske Boghandel, Copenhagen.

Mathiassen, Therkel. 1927b. *Archaeology of the Central Eskimo, II: Analytic Part.* Report of the Fifth Thule Expedition 1921–24, Vol. 4, No. 2. Gyldendalske Boghandel, Copenhagen.

Maxwell, Moreau S. 1983. A Contemporary Ethnography from the Thule Period. *Arctic Anthropology* 20(1): 79–87.

McCullough, Karen. 1989. *The Ruin Islanders: Early Thule Culture Pioneers in the Eastern High Arctic.* Archaeological Survey of Canada Mercury Series Paper No. 141. National Museum of Man Mercury Series, Ottawa.

McGhee, Robert. 1977. Ivory for the Sea Woman: The Symbolic Attributes of a Prehistoric Technology. *Canadian Journal of Archaeology* 1: 141–149.

McGhee, Robert. 1984. *The Thule Village at Brooman Point, High Arctic Canada.* Archaeological Survey of Canada Paper No. 125. National Museum of Man Mercury Series, Ottawa.

Merbs, Charles. 1983. *Patterns of Activity-Induced Pathology in a Canadian Inuit Population.* Archaeological Survey of Canada Paper No. 119. National Museum of Man Mercury Series, Ottawa.

Miller, Joshua, Eric Wald, and Patrick Druckenmiller. 2023. Shed Female Caribou Antlers Extend Records of Calving Activity on the Arctic National Wildlife Refuge by Millennia. *Frontiers in Ecology and Evolution* 10: 1059456.

Morrison, David. 1990. *Iglulualumiut Prehistory: The Lost Inuit of Franklin Bay.* Archaeological Survey of Canada Mercury Series Paper No. 142. National Museum of Man Mercury Series, Ottawa.

Oosten, Jarich. 1982. The Symbolism of the Body in Inuit Culture. *Annual for Religious Iconography* 1: 98–112.

Park, Robert. 1998. Size Counts: The Miniature Archaeology of Childhood in Inuit Societies. *Antiquity* 72: 269–281.

Petersen, Hans Christian. 1986. *Skinboats of Greenland.* Viking Ship Museum, Roskilde.

Rasmussen, Knud. 1931. *The Netsilik Eskimos: Social Life and Spiritual Culture.* Report of the Fifth Thule Expedition 1921–24, Vol. 8, Nos. 1–2. Gyldendalske Boghandel, Copenhagen.

Ray, Dorothy Jean. 1982. Reflections in Ivory. In *Inua: Spirit World of the Bering Sea Eskimo,* by William Fitzhugh and Susan Kaplan, pp. 255–267. Smithsonian Institution Press, Washington, D.C.

Savelle, James. 1987. *Collectors and Foragers: Subsistence-Settlement Systems in the Central Canadian Arctic, AD 1000–1960.* BAR International Series 358. British Archaeological Reports, Oxford.

Savelle, James, and Allen McCartney. 1988. Geographical and Temporal Variation in Thule Eskimo Subsistence Economies: A Model. *Research in Economic Anthropology* 10: 21–72.

Savelle, James, and Allen McCartney. 1994. Thule Inuit Bowhead Whaling: A Biometrical Analysis. In *Threads of Arctic Prehistory: Papers in Honour of William E. Taylor, Jr.,* edited by David Morrison and Jean-Luc Pilon, pp. 281–310. Archaeological Survey of Canada Mercury Series Paper No. 149. National Museum of Man Mercury Series, Ottawa.

Schledermann, Peter. 1975. *Thule Eskimo Prehistory of Cumberland Sound, Baffin Island, Canada.* Archaeological Survey of Canada Paper No. 38. National Museum of Man Mercury Series, Ottawa.

Schledermann, Peter, and Karen McCullough. 2003. *Late Thule Culture Developments on the Central East Coast of Ellesmere Island.* Danish Polar Center, Copenhagen.

Settle, Dionyse. 1675. *Historia Navigationis Martini Forbisseri Angli Praetoris sive Capitanei, A.C. 1577. Majo, Junio, Julio, Augusto & Septembri mensibus, Jussu Reginae Elisabethae, Ex Angliâ, in Septemtrionis & Occidentis tractum susceptae, ephemeridis sive diarii more conscripta & stilo, triennioq; post, ex gallico in latinum sermonem, à Job Thoma Freigio translata, & Noribergae, ante A. edita, denuò prodit, é muséo D. Capelli P.P.* Sumptibus Joh. Naumanni and G. Wolffii, Hamburg.

Stenton, Douglas. 1991. The Adaptive Significance of Caribou Winter Clothing for Arctic Hunter-Gatherers. *Études Inuit Studies* 15(1): 3–28.

Stewart, Andrew, Darren Keith, and Joan Scottie. 2004. Caribou Crossings and Cultural Meanings: Placing Traditional Knowledge and Archaeology in Context in an Inuit Landscape. *Journal of Archaeological Method and Theory* 11(3): 183–211.

Sturtevant, William, and David Quinn. 1989. This New Prey: Eskimos in Europe in 1567, 1576, and 1577. In *Indians and Europe: An Interdisciplinary Collection of Essays,* edited by Christian Feest, pp. 61–140. University of Nebraska Press, Lincoln.

Viveiros de Castro, Eduardo. 1998. Cosmological Deixis and Amerindian Perspectivism. *Journal of the Royal Anthropological Institute* 4(3): 469–488.

Walls, Matthew. 2012. Kayak Games and Hunting Enskilment: An Archaeological Consideration of Sports and the Situated Learning of Technical Skills. *World Archaeology* 44(2): 175–188.

Walls, Matthew. 2014. Frozen Landscapes, Dynamic Skills: An Ethnoarchaeological Study of Inuit Kayaking Enskilment and the Perception of the Environment in Greenland. PhD dissertation, Department of Anthropology, University of Toronto, Toronto.

Walls, Matthew, Pauline Knudsen, and Frederik Larsen. 2016. The Morris Bay Kayak: Analysis and Implications for Inughuit Subsistence in the Pikialarsorsuaq Region. *Arctic Anthropology* 53(1): 1–21.

Whitridge, Peter. 2001. Zen Fish: A Consideration of the Discordance Between Artifactual and Zooarchaeological Evidence for Thule Inuit Fish Use. *Journal of Anthropological Archaeology* 20(1): 3–72.

Whitridge, Peter. 2004. Whales, Harpoons, and Other Actors: Actor-Network Theory and Hunter-Gatherer Archaeology. In *Hunters and Gatherers in Theory and Archaeology,* edited by George Crothers, pp. 445–474. Center for Archaeological Investigations, Southern Illinois University, Carbondale.

Whitridge, Peter. 2013. The Imbrication of Human and Animal Paths: An Arctic Case Study. In *Relational Archaeologies: Humans, Animals, Things,* edited by Christopher Watts, pp. 228–244. Routledge, New York.

Whitridge, Peter. 2018. The Government of Dogs: Archaeological (Zoo)ontologies. In *Relational Engagements of the Indigenous Americas: Alterity, Ontology, and Shifting*

Paradigms, edited by Melissa Balthus and Sarah Baires, pp. 21–39. Lexington, Lanham, Maryland.

Whitridge, Peter. 2021. Wrapping the Body: Inuit Dolls as Fields of Real and Metaphorical Play. *Arctic Anthropology* 58(2): 218–247.

Whitridge, Peter. 2024. Manufacturing Reality: Inuit Harvesting Depictions and the Domestication of Human–Animal Relations. In *Reimagining Human–Animal Relations in the Circumpolar North,* edited by Peter Whitridge and Erica Hill, pp. 40–87. Routledge, New York.

Whitridge, Peter, and Mari Kleist. 2024. Necrontology: Housing the Dead in Precontact Labrador and Greenland. In *Exploring Ontologies of the Precontact Americas: From Individual Bodies to Bodies of Social Theory,* edited by Gordon F. M. Rakita and Maria Cecilia Lozada, pp. 15–42. University Press of Florida, Gainesville.

11

Caballito de Totora Assemblages in Ancient and Modern Huanchaco, Peru

JORDI A. RIVERA PRINCE

(*Note:* This chapter discusses human skeletal remains. There are no photos of human remains included in this chapter, but they are referenced elsewhere.)

On the shores of the Huanchaco Bay (Moche Valley, North Coast of Peru), it is impossible to miss the distinctive crescent moon–shaped reed boats that line the coast (Figure 11.1, Figure 11.2). These boats, *caballito de totora,* belong to the artisanal fishermen of Huanchaco and continue to be used as a core means to get their daily catch. This coastal fishing and surfing town is located some 11 km away from the Plaza de Armas of Trujillo and is well known internationally for its left waves, drawing tourists from around the world.

The caballito de totora are both a hallmark and symbol of Huanchaco, used in imagery like the official seal of the Municipalidad Distrital de Huanchaco, in artwork on the red buses that run from Trujillo to Huanchaco throughout the day, and in murals in the local Trujillo malls (see Rivera Prince 2023: 136 for artwork examples). The boats are made from the totora reed (*Schoenoplectus californicus* subsp. tatora plant), grown today in sunken gardens to the north of Huanchaco. The caballito de totora has been used for marine resource exploitation for over 3,500 years before present (Prieto 2016: 145), as suggested by ceramic iconography clearly depicting individuals riding a half-moon-shaped boat, a shape that very closely resembles the caballito de totora seen in modern times. Although the type of and degree of reliance on marine resources in Huanchaco has shifted from the Cupisnique (ca. 1500–1200 cal. BC) to the present day, totora fishing knowledge is still prevalent in Huanchaco and in other small, coastal towns along the northern littoral (Prieto 2016).

In the modern day, there are some 50 artisanal fishermen in the town, with about 20 of the men fishing full time (Prieto 2016: 142). With Huanchaco's modernization, there is an increasing exodus from the profession—fishermen's sons

Figure 11.1. Caballito de totora and fishing nets (green) lined up along the *malecón de Huanchaco* (Huanchaco boardwalk), north of the *muelle* (pier). Photo by Jordi A. Rivera Prince.

Figure 11.2. Caballito de totora lined up along the *malecón de Huanchaco,* north of the pier. Photo by Mahir Rahman.

Figure 11.3. Artisanal fisherman paddling a caballito de totora in the background, with tourists taking surfing lessons in the foreground, capturing the "traditional" and the "modern." Photo by Jordi A. Rivera Prince.

leave the town for university or other professional pursuits, like opening surf schools. Figure 11.3 illustrates the juxtaposition of "traditional" and "modern." Challenges to maintaining the cultural practices in the face of increasing globalization have been observed in other fishing communities (e.g., Wenzel 1995).

Throughout Huanchaco and the North Coast broadly, the caballito de totora has taken on a role beyond its use as watercraft. These boats are not just boats but are also drawn into relationships based on place, history, social interactions, and use. Beyond migration and movement, watercraft technology offered significant social changes, facilitating cultural exchange to an extent not seen before and perhaps leading to the development of social stratification and complexity (Arnold 2007; Gamble 2002). Furthermore, fishing and other marine resource extraction could occur at distances not previously achieved (or easily achieved), introducing new flora, fauna, and resources to coastal communities (e.g., Arnold 2007; Prieto 2015). Watercraft is an important tool; however, I hope to answer the call posited by Shearn (2020: 1), to move beyond travel technology and instead look to other dimensions of society, like cosmology, communal labor, and ritual dynamics that boats and watercraft play in a community.

In this chapter I move beyond typological classifications of watercraft to instead focus on the agential nature they play within their environments, with people, and with other nonhuman beings. Caballito de totora are entities themselves, in relationship with the fishermen, the sea, the marine resources, social memory, and more. While the case study and examples I detail are specific to Huanchaco and the caballito de totora, I stress that the reframing of watercraft presented here can provide coastal and maritime archaeologists an alternative, complementary way to position watercraft in their research, revealing important relationships when focus is no longer on watercraft as a technology.

The Region

Huanchaco is located on the coast of the Moche Valley, some 11 km north of the modern-day city of Trujillo. Following the descriptions offered by Pulgar Vidal (1996), Huanchaco lies in the *costa* or *chala* zone, defined by 0–500 m above sea level along the Pacific Ocean coastline, with a climate cooler than other coastlines at the same latitude. Huanchaco is therefore part of a larger Andean coastal desert punctuated by intervalley rivers that descend from the highlands—the construction of irrigation canals made plant cultivation possible along the coastline (Caramanica et al. 2020; Billman 1996, 2002).

The *chala* zone itself can be further characterized into nine resource complexes: (1) open beach sublittoral, (2) rocky shore sublittoral, (3) sandy littoral, (4) rocky littoral, (5) coastal lagoon, (6) river delta, (7) river floodplain, (8) desert, and (9) lomas, which are a seasonal fog vegetation system (Moseley 1975; see also Moseley 1992). Each complex is characterized by different water availability, flora, and fauna for exploitation (see Rivera Prince 2023: 123–24).

The fishermen of Huanchaco procure resources from three zones in the sea, detailed in Prieto (2015: 61–62); see also Rivera Prince (2023: 125–126). Sector 1 (extending ~100–200 m from the shoreline) is used for shell gathering and belongs to the community as a whole. Sector 2 (~300–500 m from the shoreline) is locally called "*el quebradero*" and is where most of the fish are caught—there are territorial divisions in this region. Sector 3 extends beyond to the open waters, and fishermen rarely go out this far.

Caballito de Totora

The caballito de totora is designed to be used by only one person at a time and is used for fishing with a line and net, and sometimes for crabbing (Gillin 1945: 35). These boats are made of an organic material and deteriorate between 12 days (Gillin 1945: 45) and less than a month (Prieto 2015: 602). When the boat is no longer usable, the fishermen unbundle the boat, save the ropes for future construction, and either leave the reeds to decompose or within the home (Prieto 2015: 602).

Each morning, the fishermen of Huanchaco rise early, head to sea, and return with their catches, which are then prepared to eat at home or sold in markets or to local restaurants. When the conditions are ideal, the fishermen leave around 4:00 to 5:00 a.m., fishing for about an hour. They go to specific places in the water depending on the time of year, the location of the sun, the type of fish they want to catch, and even sensory cues like the smell and color of the water (Prieto 2016). Artisanal fishermen recognize two seasons, one with abundant fish ("*tiempo de abundancia,*" "*la temporada,*" "*agua nueva*"), and winter, when there are higher tides (making it difficult to fish) and, due to the cold waters, fish are further offshore and catches are lower ("*friaje,*" "*la pesca chica,*" or "*el temporal,*" when "*el pescado tiene frio*") (Gillin 1945: 31, 34; Grana and Prieto 2021; Prieto 2015: 60; Rodriguez Suy-Suy 1997).

When the fishermen return to shore around 7:00 to 8:00 a.m., their families are waiting on the beach to receive them—the catch is stored in the *cal-cal* or net bag, and upon arrival the fish are washed, put into baskets, and typically taken home by the fishermen's wives (Prieto 2016). Some of the fish is distributed to others on the shore as well. The fishermen rest, then return to shore to mend their nets, lay them out to dry, and prepare for the next day around 10:00 a.m. (Prieto 2016). Around 3:00 to 4:00 p.m., the fishermen collect their drying nets, and lay out the ones to be used the next day.

Eighty years ago, fishermen's relationship to the caballito de totora was different. Both adults and children (starting at 10 years old) knew how to make the boat, although only men and sometimes young boys (playing or "surfing") were observed using the boats (Gillin 1945: 35). Today, especially with the growth of Huanchaco, this is no longer the case. While women are not known to use the boats today or in the recent past, some oral histories tell of women using the caballito de totora.[1]

Fishermen use a large geographic region beyond their homes, facilitated in part by using the caballito de totora and influenced by water temperature, food availability, salinity, and visibility (Prieto 2015: 86, 524). The caballito de totora also facilitate social interaction, with fishermen traveling to Moche, Guañape, Chao, and the Río Santa, in part due to the types of fish available in

---

1 Women were said to ride balsa rafts / caballito de totora in "El rechazo al primo del Sol" (The Rejection of the Sun's Cousin) and "Chonyicni, la princesa que se convirtió en estrella de mar" (Chonyicni, the Princess That Transformed into a Starfish) (Villadares Huamanchumo 2021: 55–58 and 75–78, respectively).

those places at certain times of the year but also due to the exchange of nets and advice with other fishermen (Gillin 1945: 34, 36).

While the use and function of the caballito de totora changes throughout time and with each generation of artisanal fishers, the role of the fishing vessel cannot be understated even today. To fully discuss and draw in the relationships that the caballito de totora engage in, it is necessary to first explore how these relationships can be framed. I stress that the particular theoretical framing I am engaging with below is rooted in Western theories and notions of process. I make this choice here as I hope this chapter inspires other archaeologists working with watercraft to consider the various processes and relationships that watercraft engage in within their own research contexts—regardless of time and place. However, in cases where local, descendant, or Indigenous knowledge and conceptualizations of watercraft and their agential nature are available, those perspectives should be given equal consideration for describing and constituting the role of watercraft in those specific areas.

Process Metaphysic, Assemblages, Practice, and Skill

Watercraft in their various forms are undoubtedly a form of technology, but as many of the other chapters in this volume highlight, they are much more than that. To illustrate the various relationships that caballito de totora are engaged in as well as their agential nature, I adopt a relational ontology to follow the various processes that are constantly in motion (see Joyce and Gillespie 2015). Defining this ontology is an important foundation for building upon the different relationships that the caballito de totora—and watercraft more generally—are constantly engaged in.

While there are many different relational ontologies, I adopt Whitehead's process metaphysic (my perspectives are further discussed in Rivera Prince and Brock Morales 2024). Important elements that distinguish Whitehead from assemblage perspectives are that his process metaphysic does not focus on nodes, meshes, meshworks, networks, and so on that often underly assemblage theories. For Whitehead, all the material objects that we can perceive in the world (and therefore interact with) are actual entities (Halewood 2005: 61; Whitehead 1978). All actual entities have virtual properties that consist of the potential ways any given entity can have meaning or interact, or both, with other entities in the world. The virtual properties become actualized, concretized, and real when they are brought into being through certain kinds of relationships (Halewood 2005: 72). I stress here that the actualization of particular virtual properties occurs with certain kinds of relationships. Actual entities then are formed as they engaged in relationships with other entities (e.g., places, people, things) throughout time. This is where I focus to understand the caballito de totora and its role in a broader assemblage. In this chapter I pay particular attention to two elements of the caballito de totora: their roles in ancient and in modern Andean history.

The Assemblage in Huanchaco

A persistent problem with assemblages is that they are infinitely large, presenting an analytical conundrum. This barrier was addressed by Delsol (2020: 195) in his theoretical framing of butchering cattle in Santiago de Guatemala, as he asked "where do we choose to set the limit to apprehend [actants] and study them?" (following Strathern 1996). Delsol defines limits to his assemblages in a multistep framing. First, drawing on Bourdieu's social fields (1979, 2013), Delsol (2020, 194) importantly highlights that these fields "are historically constructed social settings obeying a certain set of rules and accepted behaviors (norms) whereby agents compete to accumulate different forms of capital (social, cultural, economic). They can be considered as specific forms of assemblages since they relationally bind together agents, places, and objects."

However, these fields have "fuzzy" limits (Delsol 2020: 194). To further give shape to an analytical assemblage, Delsol additionally draws upon Lefèbvre's (1991: 190) "social spaces," which are "the dialectical relationship between a field (the social conditions) and the bases (the material conditions) of the action" (Delsol 2020: 195). Importantly, Delsol constitutes social spaces as the moment when these relationships take a more permanent form and it becomes possible to analyze those relationships (i.e., the assemblage).

Assemblages, however, miss an important dimension: time. In the discussions that follow, I not only focus on highlighting particular assemblages but also draw on the process metaphysic as well as perspectives on time (inspired by Lucas 2005). That is, not only do I look at archaeological constitutions of the caballito de totora but I also look to shifting examples throughout time, adding a social dimension to shift how assemblages necessarily change with new people, practices, and perspectives (i.e., relationships).

Creating a Fisher's Body

In other regions of the world, archaeologists have looked beyond watercraft themselves as evidence for their use, instead shifting focus to the "artifact assemblage" necessary for watercraft construction and maintenance (Gamble 2002: 301–302, 306), or even evidence of marine resources that could only be exploited with the use of watercraft (e.g., Arnold 2007; Prieto 2015). Here, I look beyond objects and discuss the other relationships that caballito de totora are implicated in, especially regarding the human body.

Skill

Further drawing from Delsol's framing, I also use Tim Ingold's (2001) "ecological and relational" skill to locate fishing, and the fishermen, within a greater assemblage. Skill is "a form of knowledge that draws on both the cognitive and the motor aspects of human practice that can be apprehended through a technological approach" (Delsol 2020: 196). The skill I define here is the use of these watercraft, and the social space is fishing with watercraft on the sea. As people became "enskilled" (Ingold 2001) in the practice of fishing on caballito de totora, the boats themselves were not passive; rather, by engaging in fishing and boat use, those who fished were instantiating new processes with each action, and these actions each have the effect of changing their own bodies.

José Olaya–La Iglesia Colonial de Huanchaco

The José Olaya–La Iglesia Colonial de Huanchaco site (JO-IG) is located at the highest point of Huanchaco, under the modern-day Iglesia Colonial de Huanchaco. Excavations since 2017 have occurred around the church by the Programa Arqueológico Huanchaco (PAHUAN, PI Gabriel Prieto). Inside the walls of the public primary school I.E. 80033 José Olaya Balandra are areas vacant of academic buildings, particularly to the northeast, bounded by a *canchita* (small soccer field) and an Olympic swimming pool—this is the South Sector of the JO-IG site.

Based on ceramic analyses, JO-IG had continuous occupation from the Cupisnique (Initial Period, ca. 1800–800 BC) to the Colonial period (ca. 1534–1820s) (Flores De La Oliva and Prieto 2022; Prieto 2023; Rivera Prince 2023; Rivera Prince and Prieto 2022; Villalobos Escobar 2021). My own bioarchaeological research (e.g., Rivera Prince 2023) focused on the Salinar occupation (ca. 400–200 BC in Huanchaco). For the JO-IG site, the Salinar occupation is currently understood to occur in three larger phases: (1) a cemetery, (2) "U-shaped" structures opening toward the north (facing Cerro Campana) with associated burials, and (3) continued burials with associated large ollas (or storage vessels) (see also Rivera Prince 2023: 222–225). I analyzed 84 Salinar individuals (further information on these individuals can be found in Rivera Prince 2023). Importantly, individuals ranging the life course and of all sex estimations are present in the cemetery, although the number of infants is high (Rivera Prince 2023: 372–375).

As mentioned previously, there is no current evidence of ancient caballito de totora in the North Coast of Peru, indicating that we must look for other evidence of their use in the archaeological record. As a bioarchaeologist, I am often thinking of our body's relationships to the world—not just how we shape our world but how we are shaped by it in return. Sofaer (2006) offered an interesting argument in her book, that "the body is itself created in relation to a material world that includes objects as well as other people. Throughout the life course the human skeleton may be modified through intentional or unintentional human action. During the human 'career' (Goffman 1959, 1968) bodies are literally created through social practices. The body can be regarded as a form of material culture" (Sofaer 2006: xv). So there is the potential that the material world (including caballito de totora and all the practices bundled with them) is shaped by the human body but also shaping the human body in return.

In bioarchaeology, we refer to skeletal changes associated with activity patterns as markers of occupational stress or enthesopathies. Enthesopathies are irregularities on the bone attributed to stress induced by habitual activities. For example, repetitive use of your right arm for serving a tennis ball will induce changes to the bones in your shoulder girdle over time. Elsewhere I have discussed the limitations of looking at markers of occupational stress, or enthesopathies, when analyzing human skeletal remains, which I do not detail here (Rivera Prince 2023: 456). Importantly, any given bodily movement will not necessarily have the same change to the skeleton—each person's body responds to the relationships between themselves and their environment differently based on a multitude of influences. "Factors such as age, sex, metabolic function, genetic influences, weight, and body size, all play varying roles in the presence, absence, growth, and degeneration of osteological indicators of activity" (Schrader 2019: 55). That is, the same habitual motions may not affect every individual in the same manner, or even to the same degree. To extend the previous example, one could do the same motion for serving a tennis ball to accomplish other activities, like pitching a baseball.

In observations of watching the fishermen of Huanchaco and in my lab work, an interesting pattern of skeletal changes was reminiscent of those of modern-day fishermen. I have articulated an approach joining ethnographic observation and bioarchaeology, which I refer to as "ethnobioarchaeology," to look at the potential for skeletal changes as alternative evidence for the use of caballito de totora in the past (Rivera Prince and Prieto 2019).

I start with a general description of modern artisanal fishing practices in Huanchaco to provide a foundation for articulating the bioarchaeological evidence and relationships the fishermen have with the sea and their boats. As mentioned previously, the artisanal fishermen leave for their catch early in the morning. Their caballito de totora are balanced on one shoulder to bring them to the shore so they can enter the water. Once on the boat, the fishermen kneel on the top of the reed boat with shins parallel to the boat, legs partially flexed so that their hind is elevated and the upper body erect like a kayaker (Figure 11.4 serves as a stylized example). Forward propulsion is done with a long, rectangular oar. Both carrying the boat to the water and rowing the boat produce specific stress on the body.

Elsewhere I describe the skeletal changes observed for one individual from burial IG245 (Rivera Prince 2023: 454–469). There are a number of photographs with diagrams to better illustrate the specific muscular changes observed for this individual, in addition to specific muscle attachment and origin sites expected for an artisanal fisherman using a caballito de totora. Here I describe the broad-level bodily actions required for caballito de totora use.

Consistent exposure to cold air, cold water, or both can cause exostoses in the lateral margin of the external auditory meatus. Exostoses are "typically firm, sessile, multinodular bony masses which arise from the tympanic ring of the bony portion of the external auditory canal" (DiBartolomeo 1979: 2) and develop after prolonged water exposure and irritation to the canal. Several Salinar individuals have evidence of exostoses.

The weight of the caballito de totora both entering and exiting the sea is a unilateral pressure on the body, which could potentially result in unilateral compression throughout the vertebral column or other unilateral muscular development. Leaning into the water repeatedly—for example, to lift nets with catch—could also exacerbate degenerative vertebral changes. Oar paddling would result in muscular development in the shoulder girdle, upper arms, chest, and upper back muscles. Transmission of force while rowing, particularly given the elevated position of the body, could also cause degenerative changes to the lower vertebrae, including wedging. The consistent weight from kneeling could cause stress to the patellae, which could be expressed via lipping on the patellar margins or osteoarthritic changes to the posterior patellae.

One individual in burial IG245 has nearly all of the predicted skeletal changes outlined for the use of caballito de totora. Similar changes have been reported for the nearby coastal site of Puémape, in the Jequetepeque Valley, during the formative period (ca. 2,500–1 BC) (Pezo Lanfranco and Eggers 2013). Although it is impossible to know for certain which activities caused the enthesopathies observed in this individual without direct observation of the activities, it is of note there is so much overlap in these anticipated patterns and what is observed osteologically. While future work should focus on systematic recording of these changes, there are still important implications from this discussion.

Figure 11.4. Statue on the *malecón de Huanchaco,* south of the *mulle* (pier). This statue has been in Huanchaco for over 20 years. Photo by Jordi A. Rivera Prince.

Even the potential presence of these changes suggests that fishermen do not just shape their boats but are also shaped by their boats in return. That is, the caballito de totora do not just act passively when they are used—they are not passive in this particular assemblage. Rather, the life-long skills of constructing, using, and fishing with the caballito de totora watercraft is an activity that also changes the bodies of the fishermen. Ultimately, the ongoing relationship between fishing, watercraft, the sea, and the fishermen created a "fisher's body," one permanently changed and qualitatively different from that of the rest of the community.

Other Modern Assemblages

As use of the caballito de totora endures today, it is possible to observe the different assemblages that the watercraft are a part of. Ethnographic work has broadly addressed important themes regarding modern labor divisions, environmental and policy change, and gender roles (Prieto 2015, 2016; Rivera Prince 2023). In Huanchaco, the Asociación de Pescadores Artesanales de Huanchaco–ASPAH (Association of Artisanal Fishermen of Huanchaco) is active in preserving and spreading awareness regarding caballito de totora fishing.

A great deal of time could be devoted to describing the complex and rich history and social relationships that are bundled with the caballito de totora. Here I shift focus to a few important assemblages that exist today to reveal the social entanglements that surround the caballito de totora that extend beyond the watercraft themselves. In particular, I focus on two aspects of caballito de totora: intergenerational knowledge, and crafting narratives of local identity.

These two assemblages are detailed to further highlight the agential nature of watercraft beyond their use as technology and demonstrate how relationships—assemblages or otherwise—are fundamentally intertwined with time.

Intergenerational Knowledge

Modern fishermen actively use their intergenerational knowledge, from tending to the totora reeds to the active catching of fish in the sea. Not only do fishermen tend to totora gardens but they also cut the reeds, dry them, bundle them, and construct the rafts using generational knowledge. The location of the sunken reed gardens has shifted throughout time; they are currently north of Huanchaco. While reeds are hearty plants, their care is a form of curated knowledge that endures through time. Furthermore, the construction of the caballito de totora instantiates new properties, with the most obvious being the transformation of totora reeds from reeds to caballito de totora watercraft.

Fishing itself is the instantiation of intergenerational knowledge. Not only do the fishermen know the seasonality for different marine resources but they also know the seascape itself—depths, temperatures, and currents—which all influence the types of fish that will be in certain areas at particular times. This is not to say that fishermen maintain everything the same in their practices as they did hundreds, if not thousands, of years in the past. Rather, the artisanal fishermen also change with time. As seen in Figure 11.1, the green piles—the nets—near the caballito de totora are made of nylon. While they still require care and repair for continued use, artisanal fishermen also incorporate new materials that serve similar purposes (i.e., catching fish). The incorporation of new materials within their practice creates shifting dynamics and new assemblages (e.g., new vendors, new skills for construction and repair), and the instantiation of new properties in the relationships give form to the caballito de totora.

Finally, the festivities for the Día de San Pedro highlight changing relationships and the instantiation of new properties for the caballito de totora. While the caballito de totora were used prior to Spanish colonization, based on ceramic and other iconographic representations, the introduction of Catholic beliefs highlights how new groups with new beliefs instantiate new properties for the caballito de totora. Each year for the Día de San Pedro, the fishermen of Huanchaco build a special, larger boat along with the caballito de totora called the *Patacho,* which is used for the festivities only (i.e., not to fish) (Figure 11.5, Figure 11.6). The festivities and practices surrounding the Día de San Pedro illustrate how new social dynamics (in this context, due to the forces of colonization and Spanish Inquisition) instantiate new properties for the caballito de totora—drawing them into new relationships, giving them new meanings, and drawing them into a new assemblage.

Figure 11.5. The *Patacho* for Día de San Pedro 2019. Photo by Jordi A. Rivera Prince.

Crafting Narratives of Local Identity

Finally, I briefly discuss how caballito de totora are used to craft narratives of local identity. I look at these examples to further demonstrate that caballito de totora transcend "watercraft" actively used by people beyond the artisanal fishermen and even within Huanchaco to emphasize local identity on the North Coast of Peru. Not only are caballito de totora adorning the entrance to the *mulle de Huanchaco* (Huanchaco Pier) but re-creations of the caballito de totora are put outside buildings and even bars within Huanchaco. They are even included in the official government seal of the *municipalidad de Huanchaco*. Representations of caballito de totora are used to evoke particular practices, people, and even places—further illustrating the different relationships that are constantly in motion with the caballito de totora and how their use extends beyond their role as watercraft.

One of my interests in the Trujillo area is the use of archaeological iconography in modern locations—which iconography people pick, how it is used, and where particular types of iconography are used. While future work will incorporate a more systematic analysis, here I highlight observations from my preliminary work. First, the most common iconography used in the modern

Figure 11.6. The *Patacho* for Día de San Pedro 2019, surrounded by everyday caballito de totora. Photo by Jordi A. Rivera Prince.

day in the Trujillo area is from the Moche (Middle Horizon) and Chimú (Late Intermediate Period), two cultures that had their politico-religious centers in the Moche Valley (the Huacas de Moche and Chan Chan, respectively). Local people feel a sense of connection to the Moche and Chimú peoples, who had their origins and fluorescence in the same land.

Figure 11.7. Artwork on the northwest wall of McDonald's Trujillo (Calle Francisco Pizarro 486), behind the ice cream counter with Moche art and iconography. On the left in orange and right of center in red are depictions of pescadores on a caballito de totora (marked by black arrows). Photo by Jordi A. Rivera Prince, July 15, 2023.

Interestingly, the statue in Figure 11.4 is a rare example of the caballito de totora represented with Chimú iconography (the fish in the waves). Iconography of caballito de totora is more often than not associated with Moche stylized representations. For example, Figure 11.7 is a photo of the mural on the northwest wall of the McDonald's located on Trujillo's Plaza de Armas. This McDonald's has multiple stylized representations of Moche iconography inside the restaurant, evoking a sense of Moche identity within the city center and particularly within an international (United States) fast-food chain. This mural, behind the soft-serve ice cream counter, has two different iconographic representations of fishermen on caballito de totora (see black arrows). I highlight this example given the new relationships that the boats are drawn to in this instance. Their representation is used to reinforce a sense of Moche Valley identity, creating a sense of place within an international business but with reference to an archaeological culture that existed over 1,000 years before the first McDonald's ever opened its doors.

Caballito de totora appear in communities beyond Huanchaco/Trujillo and even the Moche Valley. While caballito de totora were used beyond Huanchaco, they have today become a symbol of that place. However, the traces of the caballito de totora's previous properties and relationships are visible in other contexts. For example, in the nearby Chicama Valley (one valley to the north), the Plaza de Armas of the coastal town Puerto Malabrigo is lined with statues of caballito de totora (much larger than the boats in real life) (Figure 11.8). Although there are no longer fishermen that use caballito de totora in Puerto Malabrigo (to my knowledge), the representations of the caballito de totora are references to a bundle of practices, relationships, and properties they once held.

Watercraft as More Than Technology

As archaeologists (and anthropologists), we attempt to understand humans. However, in shifting the focal point from fishermen to the caballito de totora, it becomes possible to reveal a new set of relationships that the vessels are part of and to expand our understanding of the properties that they hold.

I advocate an intentional move beyond typological classifications of watercraft in our research, constituting boats and other forms of watercraft with a more agentive role. Such repositioning shifts away from a unidirectional dynamic in which watercraft users are the only ones engaging in practice or acting in a particular relationship. I do not argue for a "symmetrical" perspective, as is often the critique of assemblage theory (theories) (e.g., Barrett 2014 offers similar critique). Rather, I want to emphasize that important dynamics are revealed when we recognize the inherent agency of nonhuman actors (like boats or the ocean) and their agency to also instantiate change back onto people. Watercraft are not just passive, technological vessels for transport, and the sea is not just a body of water to provide food. Fishers, the sea, their communities, and watercraft globally and throughout time are relational entities acting upon one another—and constantly in states of change.

Acknowledgments

My gratitude to Colin Grier, Mikael Fauvelle, and Alberto García-Piquer for organizing the 2023 session where an earlier version of this chapter was presented, and whose suggestions made this chapter stronger. My sincere appreciation to the artisanal fishermen of Huanchaco, and PAHUAN PI Gabriel Prieto for sharing his knowledge on caballito de totora, and under whose guidance this research was completed.

Figure 11.8. Caballito de totora statues in the plaza of Puerto Malabrigo, Chicama Valley. Photo by Aleksalía Isla Alayo.

This material is based on work supported by the National Science Foundation Graduate Research Fellowship program (Award #1842473), the Ford Foundation Predoctoral Fellowship, the Wenner-Foundation Dissertation Fieldwork Grant (#10343), the Fulbright U.S. Student Program sponsored by the U.S. Department of State and Perú Fulbright Commission, and the Brown University Presidential Postdoctoral Fellowship. Any opinions, findings, and conclusions or recommendations expressed in this material are my own and do not necessarily reflect the views of the National Science Foundation, the Ford Foundation, the Wenner-Gren Foundation, the Fulbright U.S. Student Program, the government of the United States, the Perú Fulbright Commission, or Brown University.

References Cited

Arnold, Jeanne E. 2007. Credit Where Credit Is Due: The History of the Chumash Ocean-going Plank Canoe. *American Antiquity* 72(2): 196–209.

Barrett, John C. 2014. The Material Constitution of Humanness. *Archaeological Dialogues* 21(1): 65–74.

Billman, Brian R. 1996. The Evolution of Prehistoric Political Organizations in the Moche Valley. PhD dissertation, Department of Anthropology, University of California Santa Barbara, Santa Barbara.

Billman, Brian R. 2002. Irrigation and the Origins of the Southern Moche State on the North Coast of Peru. *Latin American Antiquity* 13(4): 371–400.

Bourdieu, Pierre. 1979. *La Distinction: Critique sociale du jugement* [Distinction: A social critique of the judgment of taste]. Éditions de Minuit, Paris.

Bourdieu, Pierre. 2013. Séminaires sur le concept de champ, 1972–1975 [Seminars on the concept of field, 1972–1975]. *Actes de la recherche en sciences sociales* 5(200): 4–37.

Caramanica, Ari, Luis Huaman Mesia, Claudia R. Morales, Gary Huckleberry, Luis Jaime Castillo Butters, and Jeffrey Quilter. 2020. El Niño resilience farming on the north coast of Peru. *Proceedings of the National Academy of Sciences of the United States of America* 117(39): 24127–24137.

Delsol, Nicolas. 2020. Disassembling Cattle and Enskilling Subjectivities: Butchering Techniques and the Emergence of New Colonial Subjects in Santiago de Guatemala. *Journal of Social Archaeology* 20(2): 189–213.

DiBartolomeo, Joseph R. 1979. Exostoses of the External Auditory Canal. *Annals of Otology, Rhinology & Laryngology* 88(6 suppl.): 2–20. https://doi.org/10.1177/00034894790880S601.

Flores De La Oliva, Luis, and Gabriel Prieto. 2022. Dinámicas ocupacionales entre el Horizonte Temprano Tardío y el Periodo Intermedio Tardío en la bahía de Huanchaco: Una perspectiva del área 27, del Sector José Olaya, sitio Iglesia Colonial de Huanchaco. *Actas del Congreso Nacional de Arqueología* 7: 393–406.

Gamble, Lynn H. 2002. Archaeological Evidence for the Origin of the Plank Canoe in North America. *American Antiquity* 67(2): 301–315.

Gillin, John. 1945. *Moche: A Peruvian Coastal Community.* Smithsonian Institution Institute of Social Anthropology Publication No. 3. Smithsonian Institution, Institute of Social Anthropology, Washington, D.C.

Goffman, Erving. 1959. *The Presentation of Self in Everyday Life.* Penguin, Harmondsworth.

Goffman, Erving. 1968. *Asylums: Essays on the Social Situation of Mental Patients and Other Inmates.* Penguin, Harmondsworth.

Grana, Lorena, and Gabriel Prieto. 2021. Marine Diatom Remains as Bioindicators of the Uses of Pre-Hispanic Fishing Gear Recovered in Ritual Contexts at Huanchaco, North Coast of Peru. *Journal of Archaeological Science: Reports* 39: 103167.

Halewood, Michael. 2005. On Whitehead and Deleuze: The Process of Materiality. *Configurations* 13(1): 57–76.

Ingold, Tim. 2001. Beyond Art and Technology: The Anthropology of Skill. In *Anthropological Perspectives on Technology,* edited by Michael B. Schiffer, pp. 17–31. University of New Mexico Press, Albuquerque.

Joyce, Rosemary A., and Susan D. Gillespie. 2015. *Things in Motion: Object Itineraries in Anthropological Practice.* School for Advanced Research Press, Santa Fe.

Lefèbvre, Henri. 1991. *The Production of Space.* Translated by Donald Nicholson-Smith. Blackwell, Oxford, UK.

Lucas, Gavin. 2005. *The Archaeology of Time.* Routledge, London.

Moseley, Michael E. 1975. *The Maritime Foundations of Andean Civilization.* Cummings Press, Menlo Park.

Moseley, Michael E. 1992. Maritime Foundations and Multilinear Evolution: Retrospective and Prospective. *Andean Past* 3:5–42.

Pezo Lanfranco, Luis Nicanor, and Sabine Eggers. 2013. Modo de vida y expectativas de salud en poblaciones del periodo formativo de la Costa Norte del Peru: Evidencias bioantropológicas del Sitio Puémape. *Latin American Antiquity* 24(2): 191–216.

Prieto, Gabriel. 2015. Gramalote: Domestic Life, Economy and Ritual Practices of a Prehispanic Maritime Community. PhD dissertation, Department of Anthropology, Yale University, New Haven, Connecticut.

Prieto, Gabriel. 2016. Balsas de totora en la costa norte del Perú: Una aproximación etnográfica y arqueológica. *Quingnam* 2: 141–188.

Prieto, Gabriel. 2023. La ocupación del Horizonte Temprano Tardío (400–200 cal. a.C.) en Huanchaco: Vida Cotidiana y Prácticas Ceremoniales. *Arqueología y Sociedad 39*: 53–86.

Pulgar Vidal, Javier. 1996. *Geografía del Perú: Las Ocho Regiones Naturales, la Regionalización Transversal, la Sabiduría Ecológica Tradicional.* 10th ed. Promoción Editorial Inca S.A. (PEISA), Lima.

Rivera Prince, Jordi A. 2023. Emerging Inequality: Investigating Social Differences in Ancient Huanchaco, North Coast of Peru (400–200 cal. BC) Through Bioarchaeological and Mortuary Methods. PhD dissertation, Department of Anthropology, University of Florida, Gainesville.

Rivera Prince, Jordi A., and Amanda Brock Morales. 2024. Backdirting: Theorizing Backdirt Through Time, Place, and Process. *Journal of Field Archaeology* 49(2): 98–114.

Rivera Prince, Jordi A., and Gabriel Prieto. 2019. Defining Markers of Occupational Stress in the Ancient Fisherman of Huanchaco, Peru: When Modern Ethnography and Bioarchaeology Intersect. Paper presented at the 84th Annual Meeting of the Society for American Archaeology, Albuquerque, New Mexico.

Rivera Prince, Jordi A., and Gabriel Prieto. 2022. Primeras aproximaciones bioarqueológicas de una muestra de tumbas Salinar (400–200 cal. a.C.) en la Bahía de Huanchaco: Propuestas y Perspectivas. *Actas del Congreso Nacional de Arqueología* 7: 621–631.

Rodriguez Suy-Suy, Antonio. 1997. *Los Pueblos Muchik en el Mundo Andino de Ayer y Siempre.* Centro de Investigacion y Promocion de los Pueblos Muchik "Josefa Suy Suy Azabache," Moche, Perú.

Schrader, Sarah. 2019. Bioarchaeological Approaches to Activity Reconstruction. In *Activity, Diet and Social Practice: Addressing Everyday Life in Human Skeletal Remains,* pp. 55–126. Bioarchaeology and Social Theory. Springer International, Cham.

Shearn, Isaac. 2020. Canoe Societies in the Caribbean: Ethnography, Archaeology, and Ecology of Precolonial Canoe Manufacturing and Voyaging. *Journal of Anthropological Archaeology* 57:101140.

Sofaer, Joanna R. 2006. *The Body as Material Culture: A Theoretical Osteoarchaeology.* Cambridge University Press, Cambridge.

Strathern, Marilyn. 1996. Cutting the Network. *Journal of the Royal Anthropological Institute* 2(3): 517–535.

Villadares Huamanchumo, Percy. 2021. *Historias del Abuelo.* Ediciones Rafael Valdez E.I.R.L. and Institute of Andean Research, Lima and New Haven.

Villalobos Escobar, Antonio. 2021. Organización social y ritual funerario en los alrededores de La Iglesia Colonial de Huanchaco durante la ocupación Salinar. Bachelor's thesis, Universidad Nacional de Trujillo, Trujillo, Perú.

Wenzel, George W. 1995. Ningiqtuq: Resource Sharing and Generalized Reciprocity in Clyde River, Nunavut. *Arctic Anthropology* 32(2): 43–60.

Whitehead, Alfred North. 1978. *Process and Reality: An Essay in Cosmology,* rev. ed. Edited by David Ray Griffin and Donald W. Sherburne. Free Press, New York.

12

Toward, Not To

Seafaring Worldviews from Viking Age and High Medieval Norway

GREER JARRETT

For the coastal communities of Scandinavia, seaborne movement has been a ubiquitous aspect of life for at least the last four millennia (Østmo 2020). Multidisciplinary evidence for ancient maritime activity connects people, places, objects, and ideas across this region's seascapes but rarely provides us with more than a point of origin and eventual destination, deposition, or burial (Marcus 1980: 108). Surviving information about the sea voyages between these points is extremely scarce until the development of maritime ethnology in this region in the early- and mid-twentieth century (Færøyvik and Færøyvik 1979; Hasslöf et al. 1972; Westerdahl 1989; Figure 12.1). This scarcity hinders our understanding of the practical aspects of ancient maritime mobility (navigation and sailing, routes and itineraries, voyage duration, seasonality, costs and risks) as well as the cognitive and ontological dimensions of seaborne travel. In this chapter, I address this problem by exploring how human mobility patterns are entangled in culturally specific worldviews and how reconstructing the worldviews of ancient seafaring societies may help us retrace and characterize their voyages.

Seafaring is a skilled, collective, and intrepid form of maritime activity (Broodbank 2006: 200, 208) that played a central role in Scandinavia during the Viking Age (ca. AD 800–ca. 1050) and the High Middle Ages (ca. AD 1050–1300) (Bill 2010: 20). Over this timespan, Scandinavian maritime communities gradually lost most (but not all) of the attributes that would characterize them as small-scale societies, such as low population densities, subsistence economies, decentralized social and political organization, and religious diversity (Reyes-García et al. 2017; for a discussion of this in relation to the Norwegian Iron Age, see Berthelsen 1997). This loss also incurred the shedding and ac-

Figure 12.1. Bernhard Færøyvik measuring a larger Åfjordsbåt known as a *fembøring* in 1936. The thorough documentation work conducted by Færøyvik and others during this period allowed for the survival of the Åfjord tradition and informed the construction of the boats used in this study. Photograph by Kristian Kielland, Norwegian Maritime Museum, Id: NSM.1701-161.

tive suppression of a pre-Christian and pre-urban maritime worldview, the reconstruction of which forms the focus of this chapter. The primary material for this reconstruction is made up of a series of experimental voyages on board traditional Norwegian boats, cultural and technological descendants of the boatbuilding and sailing traditions of the Viking Age (Figure 12.4). Through an analysis of these voyages, I identify the primary affordances of route choice and navigation and present the likely characteristics of the accompanying worldview. By better understanding the emergent relationship between seafaring practices and worldviews, it becomes possible to get a sense of the suite of risks and opportunities that Viking Age and high medieval sailors would have perceived during their voyages and thus suggests which routes they would have favored.

Like the other authors of this volume, during this research I have attempted to give voice to the perspective of the sailors who undertook these voyages, and to "take seafaring on its own terms" (Grier et al., Chapter 13, this volume). Such a perspective is often lacking in our profoundly terrestrial academic tradition, making this attempt of use to other scholars also seeking to "think from the ocean" (Steinberg and Peters 2015: 261, as well as most of the chapters in this volume). As some have shown, our lack of knowledge about ancient voyages (by which I mean physical and planned seaborne movement using culturally

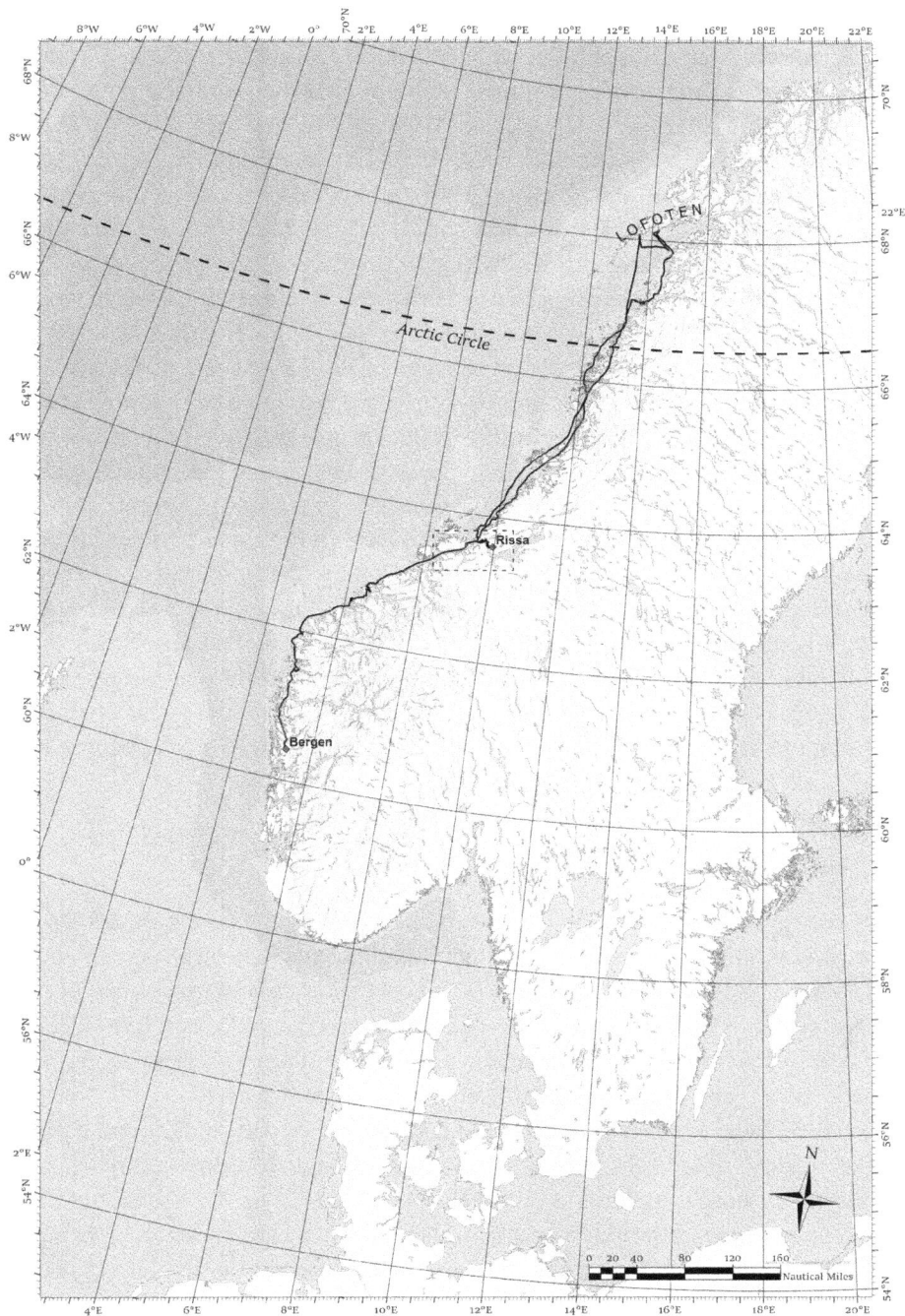

Figure 12.2. The Norðvegr, with places mentioned in the text. The lines indicate the return voyage from Rissa to Lofoten and the voyage from Rissa to Bergen, and the rectangle marks the area where the sailing trials were conducted.

specific watercraft) cannot be exclusively blamed on a lack of evidence: we are at least equally impeded by the great ontological distance separating modern scholar from ancient sailor (Frog et al. 2019: 13; Safadi and Sturt 2019: 1). This distance seems to originate in the context of use—that is, in the way culturally specific activities and environments offer up distinct understandings of space, movement, and the world (Hutchins 1995; Ingold 2000). To bridge the gap between different contexts of use, studies of maritime mobility face the challenging task of developing methods that can acknowledge, analyze, and interpret past maritime activities and environments for a modern and terrestrial audience. This means actively moving beyond the conventional analytical and representational tools of western science, which often encourage the reduction of ancient voyages to abstract values or colored arrows across empty blue expanses (Edney 1999: 167; Gillings 2012: 608; recent examples of such reductions can be found in Brink and Price 2012; Cunliffe 2017; Østmo 2020). To gain a fuller understanding of Viking Age and high medieval seafaring (which includes voyages but also the technical knowledge, skills, and experience required to undertake them), we therefore need alternative approaches and models at least as much as we need new evidence. As several of the chapters in this volume exemplify, creative and critical approaches to ancient seafaring are in a stage of healthy development in the Americas (e.g., García-Piquer, Chapter 2; Rorabaugh, Chapter 3), but a notable lag in this innovation is apparent in the context of Scandinavian archaeology.

Seafaring as Wayfaring

In exploring these alternatives, I have found a strong alignment between new breakthroughs in neuroscience and theoretical perspectives on wayfaring and affordance. Both suggest that human mobility is culturally and environmentally contingent, and that an understanding of the context of use is therefore essential for reconstructing ancient pathways. A brief outline of these will serve to preface their application to the project's experimental voyages.

We can begin with neuroscientific understandings of path-finding and navigation. According to the current consensus, spatial and directional information from sensory experience and path integration is gathered, computed, and represented in the hippocampus. This results in a regular pattern of firing neurons, often referred to as a cognitive or mental map, due to its role in spatial awareness and navigation (Hafting et al. 2005; Tolman 1948; Whittington et al. 2022). However, recent research on entorhinal grid cells, which play a key role in this process, suggests that we should not think of the information these cells encode as a direct mental reproduction of "what is out there." Rather than "maps," these cells create structured frameworks of cognition for organizing and communicating thoughts and experiences about wayfinding and navigation but also about family trees, social networks, or sequential practices (Whittington et al. 2022). For movement and navigation, this framework serves as a kind of metronome of abstraction, allowing the traveler to integrate spatiotemporally encoded impressions of their surroundings into a broader understanding of space in a consistent fashion, thereby revealing possible routes or itineraries toward an intended goal or destination.

It is the yielding of potential pathways through relational thinking that makes cognitive processes so crucial for studying ancient mobility. Significantly, and contrary to earlier hypotheses, grid cells seem to be strongly determined by particular environmental cues, meaning that the patterns they create do not exist independently of our experience and movement through our surroundings (Lisman et al. 2017; cf. Hafting et al. 2005). The neuroscientific shift toward context and dynamism aligns remarkably well with relational readings of Gibson's theory of affordances (Chemero 2003; Gibson 2014: 119–135) as well as archaeological applications thereof (Gillings 2012; Wernke et al. 2017). According to these, a community's shared movements, practices, and environments shape their perception, understanding, and representation of the world. This encourages and constrains certain patterns of choice and action, which in turn generate particular conceptions of space and particular ways of finding one's way through it (Ingold 2000: 161; 2011: 151). Thus, both scientific and theoretical perspectives seem to reject the static, universalizing metaphor of the mental map and encourage us to imagine ancient paths or routes as being afforded by particular worldviews, which originate, in turn, in patterns of movement and practice through particular environments.

It follows that any reconstruction of ancient maritime mobility must begin with an understanding of the affording worldview. Central to this chapter's approach is the ethnoarchaeological hypothesis that such an understanding can be reached through active, long-term engagement in the practices and environments under study, in collaboration with the local bearers of descendant traditions (David and Kramer 2001: 59; Lane 2016: 605). This engagement allows the fieldworker to inhabit a common space of experience with ancient practitioners and thereby attune themselves to a similar suite of affordances. The establishment of this experiential nexus encourages the development of similar understandings of space and movement, allowing the modern scholar to suggest which routes might have been favored in the past (Ingold 2000: 166–167, 216).

Fortunately, practical engagement in traditional Scandinavian seafaring has a long and decorated history, with over a century of research by archaeologists and ethnologists. Experimental sailing voyages on board historical and reconstructed boats and ships have revitalized expertise in ancient sailing techniques and resulted in a relatively standardized data-gathering method, which is followed in this study (Bischoff et al. 2014; Crumlin-Pedersen et al. 1980; Englert 2006, 2012). Parallel to these efforts, ethnological research among traditional seafaring communities across Scandinavia has revealed strong continuities over centuries or even millennia in the practices of boatbuilders, sailors, and fisherfolk, suggesting that fragments of a possibly pre-Christian worldview may have also endured (Eldjárn and Godal 1988: 17; Westerdahl 2010).

The insights gained from experimental archaeology and ethnology illustrate the potential of alternative approaches to ancient maritime mobility and highlight the importance of physical engagement when studying fundamentally practical traditions such as navigation, sailing, and boatbuilding. This study expands on previous research by focusing on sailing voyages along the Norðvegr (the coastal route along the western coast of the Scandinavian Peninsula; see Figure 12.2), an area that has received less scholarly attention than southern Scandinavia (e.g., Englert and Trakadas 2009; Indruszewski and Godal 2006). The reconstructed ontology draws on voyages conducted by the author as well as previous scholarship regarding seafaring practices and worldviews along this coastline (e.g., Storli 2007; Valtonen 2008). The study period (AD 800–1300) spans the time from the first indisputable evidence for the use of clinker-built sailing boats like those used in the project's fieldwork to the diffusion of the magnetic compass and the nautical chart, which afforded abstract and increasingly modern conceptions of space (Kemp and D'Olier 2016).

Experimental Methods

This project's primary data-gathering method consisted of long-term, practical engagement in traditional sailing practices along the Norwegian coast and the consequent establishment of an experiential nexus with sailors from the past, as discussed above. The main objective was to identify the affordances of maritime mobility and route choice that were revealed through these shared experiences. The affordances were then compared with historical and archaeological evidence for their presence in the timespan under study and used to reconstruct the characteristics of a Viking Age and high medieval seafaring ontology, which is presented at the end of the chapter (Figure 12.3).

The fieldwork consisted of a series of sailing trials and trial voyages on board six different Norwegian boats built in the Åfjord tradition of the late nineteenth century (Figure 12.4). For the purposes of this study, it seemed most fruitful to use vessels whose handling and constituent parts are fully evidenced and understood, rather than adding more uncertainty by reconstructing an older,

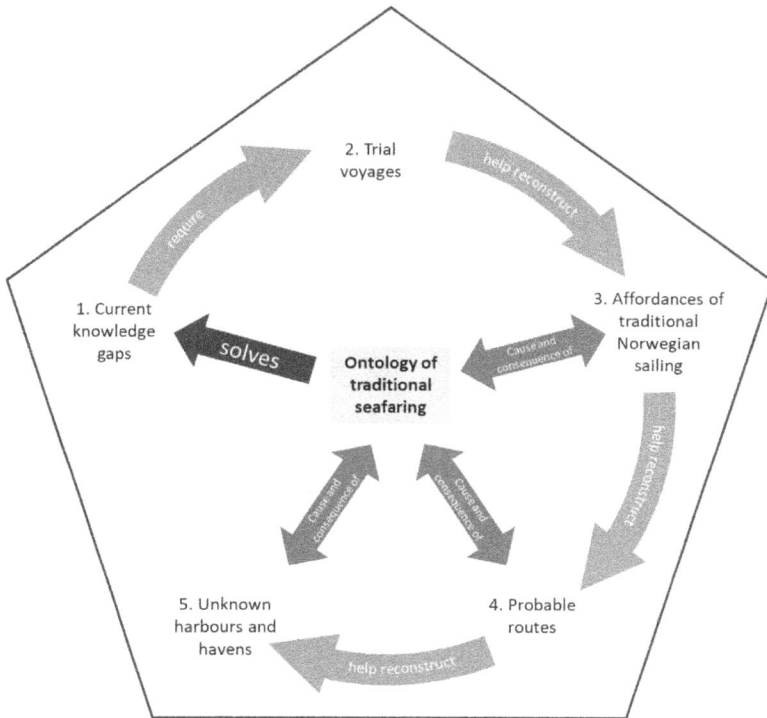

Figure 12.3. This chapter's methodology in schematic form. Note the two-way relationship between worldview and practice.

less well-preserved, and more hypothetical craft. Åfjord boats are double-ended, square-rigged vessels; clinker-built of pine, spruce, or both; and equipped with multiple pairs of oars. They were used until the early twentieth century for fishing and transport throughout Trøndelag but are most renowned for their annual voyages to and from the cod spawning grounds off the Lofoten archipelago, a round trip of some 700 nautical miles typically undertaken between January and April (Eldjárn and Godal 1990; Parsons 2013). The boats' basic characteristics are common to Norwegian boatbuilding for at least the last 1,200 years; in this sense, late examples of the tradition can serve as analogies for exploring what routes might have been followed along the Norðvegr in earlier periods (Eldjárn and Godal 1988: 13; Weski 2006: 64). The uninterrupted continuity of this heritage, coupled with the role these craft played in both local and regional mobility networks, makes them particularly promising candidates for the present research.

Figure 12.4. A traditional Norwegian boat built following the Åfjord tradition. Rissa, Trøndelag, September 2021. Just like the boats of the Viking Age, Åfjord boats are built frame-first and rigged with a square sail. Image courtesy of Tora Heide.

Fifteen sailing trials and two trial voyages were undertaken between September 2021 and June 2022. The sailing trials, lasting between one and four days, were conducted in and around the Trondheim Fjord in a wide range of weather conditions and in all interceding months. Two longer voyages of approximately three weeks each were conducted between April and June 2022: the first was a return voyage from Rissa to Lofoten, and the second a one-way voyage from Rissa to Bergen (see Figure 12.2). A total of 1,494 nautical miles were covered over ground during this campaign.[1] The sailing trials were used as a way for the volunteer crews to gain familiarity and skill in the practices of sailing as well as to allow for the study of specific aspects of the vessels' performance, such as the possibility of recovering after a capsize or the boats' windward sailing capabilities. The two voyages were conducted after the trials and focused on evaluating potential ancient routes and havens along the Norðvegr.

Data was gathered before and during these trials in the form of field notes, interviews, photographic and video footage, meteorological data and obser-

---

1 Two of the six Åfjord boats from this project were equipped with engines, which were used on two occasions, and there were four instances of towing; this amounted to about 15 hours, or 3.1%, of the 467 hours spent in motion.

vations, and a GPS track of each trial. This provided a comprehensive record, including the performance of the various vessels under a range of conditions, the routes taken, and the range of human and environmental affordances of mobility and route choice. Upon completion of each voyage, field notes, photographs and other data were organized and digitized, and an evaluation was made of which factors were the primary affordances of the route chosen.

Results

What follows is a presentation of the primary affordances of route choice identified during the project's sailing trials and trial voyages. These results inform the seafaring ontology outlined at the end of this chapter as well as the proposed reconstructions of Viking Age and medieval sailing routes, which are elsewhere (Jarrett 2025; Ruiz-Puerta et al. 2024).

When traveling aboard an open boat without an engine and powered only by wind and oars, the foremost contrast with modern travel is the conception of travel itself (discussed in the context of Indigenous Baja Californian seafaring by Des Lauriers and García-Des Lauriers, Chapter 5, this volume). In an interview with skipper and sailing teacher Lena Börjesson, I asked how she would think of a sailing voyage to a specified destination both before and during a voyage. She answered: "Jag vill inte segla *till* någonting . . . , jag vill segla *mot*" ("I do not want to sail *to* somewhere, I want to sail *toward* it"; personal communication 2022). This statement illustrates the inherent unpredictability of a voyage under sail: in October 2021, during a trial in the Trondheim Fjord, our intended destination changed no less than four times within one day. Even with accumulated experience, such changes remained common the following year; during an attempted crossing of the Vestfjord toward Lofoten in May 2022, major wind changes over a 24-hour period forced us first to abandon the crossing and then to entirely change our itinerary, arriving the following morning in Bodø, some 50 nautical miles southeast of our intended destination.

The opportunistic and adaptable sailing practices evident in these experiences are often presented in the academic literature as the harnessing of "favorable" weather conditions, but what this means is rarely explained (e.g., Englert 2012: 273; Marcus 1980: 101). During this study it became clear that "favorable" is rarely synonymous with "pleasant" or "optimal." On the first of May 2022 we sailed from Leka toward Sandnessjøen, covering 65 nautical miles in just under 12 hours despite constant heavy rain and fog and winds up to Beaufort 7 (near gale), requiring one and sometimes two reefs in the mainsail. In contrast, a second crossing of the Vestfjord was attempted on the eighth of May from Bodø in what first appeared to be "favorable" weather, with calmer

Figure 12.5. Bremangerlandet, June 2022. After a successful passage around Stad in large but gentle swells, we encountered strong katabatic gusts and steep waves upon approaching land. This experience highlighted the risk of approach areas even when conditions are favorable on the open sea.

winds (Beaufort 2–4), clear skies, and excellent visibility. However, conditions throughout the morning became very unstable, with snow flurries, sleet, and gale-force gusts descending from Landegode. The skipper therefore decided to abort the crossing once again, and we changed course for Steigen. As these episodes show, the stability and predictability of the weather is much more conducive to long-distance sailing than seemingly optimal conditions in the present moment (Figure 12.6).

Spells of stable, usable weather represent opportunities that cannot be missed, regardless of when they occur. This sometimes involves sailing through the night, and the trials proved that this is perfectly achievable between May and July along the Norwegian coast (see Figure 12.6), as has also been shown for the Baltic by Indruszewski and colleagues (2008). What is perhaps less clear in current scholarship is the relative danger posed by different stretches and features of the Norðvegr. Many scholars have highlighted the risks of exposed areas such as Hustadvika or Stad (e.g., Kruse 2017; Østmo 2020; Skre 2014). However, during this study the moments of greatest peril occurred along the relatively protected approaches to and from exposed areas, such as in Linesf-jorden, Breisundet, or along the south side of Bremangerlandet (Figure 12.5). Narrow sounds or coastal waters featuring sharp bathymetric change can be dangerous even when the weather and wind conditions out to sea are stable and favorable. Irregular, steep wave patterns, strong tidal currents, reefs, and the

Figure 12.6. *Top:* A *fembøring* sailing at night, May 2022. *Middle:* A clear example of unstable weather, Sandnessjøen, April 2022. *Bottom:* A world within a world: the profoundly social nature of life on board. Trondheimsleia, May 2022.

infamous katabatic winds known as *fallvind* constitute major risks for open sailing boats that are often disregarded in academic literature (Eldjárn and Godal 1988: 94). In a discussion about possible routes during the trial voyage from the Trondheim Fjord to Lofoten, skipper Kjetil Sildnes described a route running from Buholmråsa to Vega via Ytter-Vikna, Hortavær, and Muddværet that he used when the wind blew from the east or southeast (personal communication 2022). This route deliberately kept away from the coast to avoid the *fallvind* and take advantage of stronger and steadier winds further out to sea. This evidence suggests that sailing crews may have needed to wait for safer conditions not only when departing for the open sea but also when approaching the mainland, indicating that ancient maritime nodes may have been located along the outer coast rather than within the inner fjords, as they are today.

The subject of navigation in Viking Age Scandinavia has generated decades of debate (for a comprehensive recent discussion, see Filipowiak 2020). Although safety concerns meant that the boats used during these trials were equipped with some modern navigational aids (paper charts, compass, and VHF radio), some remarks can still be made on this topic. The Norðvegr is primarily a coastal environment flanked by monumental landforms with unique associated names, poems, and tales (Morcken 1978; Westerdahl 2010). During the trials this allowed for a combination of pilotage using named land- and seamarks (such as Skrova, Bolga, Landegode, and Stemshesten) and careful dead reckoning during open-sea stretches or in low visibility (sometimes using a tally system, as suggested by Marcus [1980: 108]). In the absence of digital navigational instruments displaying absolute data, extreme attentiveness and effective communication had to be maintained during critical navigational episodes between the lookout, the navigator, and the person steering. Collective attentiveness was particularly important because much of the information used for making decisions about routes or destinations was available to all crew members in their surrounding environment (landmarks, wind direction, sea state, incoming weather systems). This meant that many of the skipper's decisions were made collectively, or at least in consultation with other members of the crew, rather than through the reading of specialized instruments displaying absolute data.

Navigation was by no means the only collective task, however. The trials involved crews of 4–15 people, and although rotating roles were usually assigned, a great deal of the work was undertaken collectively, such as reefing, tacking, mooring, or preparing food. To be able to sleep while under way, the off-duty watch needed to have total trust in the technical competence of the watch on duty. This collective reliance led to the development of strong social bonds during the sailing trials, allowing the skipper to delegate an increasing level of responsibility to the crew, and for the crew to develop confidence in themselves. Social networks also extended beyond the boat and included people encountered or accounted for along the way, who played an important role in providing mooring and shelter, local knowledge and information, and access to supplies or materials. It was thanks to such contacts that we heard of the unnatural calms around Hitra (Kenneth Bjørkli, personal communication 2021), the dangers of the Breisund (crew of *Storeggen*, personal communication 2022), and the hidden channel south of Askvoll known as Sauesund (Arvid Thuland, personal communication 2022).

This collective, social dimension is arguably the most dominant of the seafarer's experiences. The interactions, relationships, and dynamics between the people involved in the voyage are of primary consequence for the voyage's route, nature, duration, and outcome (see Figure 12.6). Additionally, collective tasks and the total absence of privacy meant that technical skill and experience did not always matter most; the crew's morale, their capacity for teamwork, and their levels of mutual trust were equally crucial.

Attunement to the Seascape through Practice at Sea

The sailing trials involved repeated routes along and around the Trondheim Fjord. As the same seascapes were repeatedly traversed, previously unperceived or unknown elements of practice and of the environment became deeply familiar. This familiarity developed collectively as the crew learned to work and communicate together using shared skills, place-names, and nautical terminology. The process and extent of this experiential transformation cannot be described in its entirety here, but several of its effects are significant for the present discussion.

First, measurements of absolute distance were rapidly shed in favor of notions of voyage duration. This was apparent already in October and November 2021, probably due to the hard lessons learned in the mercurial autumn weather; questions like "how far is it?" rapidly became "how long do you reckon it will take?" Throughout the following months, the crews gradually came to trust their own rapidly expanding experience over the information provided in textual sources, such as tide tables, as these rarely accounted for complex local conditions.

Second, as notions of absolute distance faded, a network of known and named landmarks slowly developed in the collective geography on board. We began to think of particular routes not as lines on a map but as a sequence of remembered places colored by the accompanying experience of sailing in their vicinity. The places where several possible routes met, such as Agdenes, Bolga, or Landegode, became particularly important mnemonic anchors, as it was here that final choices had to be made. Thus, the crew began describing routes as a sequence of ordered place-names: typically this would be phrased as "you sail toward [place A], and then when you pass it you have [place B] to port/starboard, and then you have [place C], and after that [place D]." Few details are given about these places, other than perhaps the most serious dangers, as it is assumed the listener will know of them. During the trials, we continued to use nautical charts for navigation, but this kind of embodied remembering of a route, and of the good or bad choices made, often outshone the mental image of a plotted course to such a degree as to almost replace it.

The third and final effect to mention here is related to the development of practical skills and bonds of mutual trust. These brought about a rapid expansion in the crew's conception of what routes and conditions were deemed safe, opening up the seascape to possible voyages that would previously not have been considered. For example, initial encounters with katabatic winds, winter sailing, or rigging breakages led to changes or abandonment of attempted sailing trials. But having learned from these encounters, similar experiences during the trial voyages were no longer perceived as seriously dangerous, allowing us to continue toward our intended destination. Conducting experimental trials in suboptimal conditions was thus of huge benefit to the general results of the project as the crews developed a level of competence that would otherwise never have been reached.

Navocentrism, Attendance, and Perceived Risk

As such sequential descriptions make clear, the orientation system at work during this kind of sailing is entirely centered on the moving boat. This perspective is reinforced by the total dynamism of the surrounding land- and seascape as it seemingly rises, falls, and revolves around the observer and appears and disappears behind curtains of weather, rolling waves, or other landforms. The boat thus becomes the only fixed axis of orientation, with places and possible routes considered in relation to it and its movement along a remembered path. As the trials primarily involved coastal navigation, this orientation system took mental precedence over modern abstract geographies of regions beyond the horizon. It would be of great interest, however, to see whether traditional sailing practices afford this kind of "navocentric" orientation system even during voyages out of sight of land.

One of the primary results from the experimental trials was the establishment of a clear relationship of affordance between the dynamism of the environment and the nature of traditional Norwegian navigation and seafaring practices. The changing and unpredictable nature of the weather and the sea along the Norðvegr required constant and collective attentiveness from the crew, as any change, no matter how small, could herald a vital opportunity or the advent of disaster. Navigational choices were made according to that which was collectively perceived, known, or believed to exist in the environment. This implied a constant level of uncertainty: decisions and choices taken were rarely objective calculations based on quantitative data but rather were judgments of possible outcomes based on perception, knowledge, and experience. An example of this was our changing attitude to potentially dangerous areas: early encounters with the strong currents around Agdenes, a promontory that features frequently in the medieval sagas (Sturlason 1990: 153, 373, 559), led us to give this headland a wide berth during the initial sailing trials. But as the crew became familiar with the tidal cycle in this area, we learned to judge when it was safe to take the more direct route close to the promontory. Although uncertainty remained, our accumulated experience changed our attitudes to Agdenes from an area to be categorically avoided to a place whose potential risks could be judged through attendance to our increasingly meaningful surroundings.

To return to this chapter's main research question, the central factor that afforded route choices during the project's trials were these value judgments of possible outcomes—in other words, judgments of perceived risk. This is perhaps best expressed as the following question, rarely voiced but clearly implicit in the choices made during these trials: Judging by the current circumstances, does taking this route represent an acceptable level of risk to vessel and crew?

Seafaring Affordances Through Time

The traditional boats and sailing practices studied during the project's trials encouraged the development of an undeniably distinct conception of maritime space and movement, anchored in collective judgments of perceived risk. However, the value of this as an analogy for studying ancient seafaring depends on the identification of similar patterns of practice and movement through similar environments in the past. Here I briefly present parallels between the seafaring affordances identified during this project and those apparent in the surviving evidence for maritime mobility along the Norðvegr in the Viking Age and High Middle Ages.

The contingent nature of traditional sailing is highlighted both in modern scholarship (Campbell 2020) and in medieval written sources. In *The King's Mirror*, for example, the narrator describes fair weather as a temporary peace between dangerous forces (Larson 1917: 90). In Norwegian folklore, these dangers are personified in the figure of the *draug* (Mathisen and Sæther 2018). The use of watches to allow for night sailing during this temporary peace is attested for in the archaeological evidence for Viking Age ships (Ellmers 1995: 237; Ravn 2016: 133) and in later saga literature (e.g., Jónsson 1947: 324).

The core elements of traditional Norwegian boatbuilding and sailing practices display high levels of continuity across both space and time. In their study of boatbuilding from Åfjord and Nordland, Eldjárn and Godal (1990: 79) were able to identify a system of proportional measurement used from the time of the Gokstad ship until the early twentieth century. Perhaps even more strikingly, Parsons (2013) has shown that many of the terms, roles, practices, and even knots recorded by these two authors in the mid-twentieth century were in use on board West Highland galleys in medieval Scotland, suggesting that these traditions accompanied the introduction of Scandinavian boatbuilding to the Celtic world during the Viking Age while remaining in continuous use in Scandinavia ever since.

With similar boats, the danger of approach areas was also a historical constant, as is expressed regularly in the sagas (e.g., Smiley and Kellogg 2001: 54; Turville-Petre and Olszewska 1942: 58). The need for outlying maritime nodes mentioned above may also explain the location of the five earliest Norwegian royal manors linked to Harald Fairhair, which are closer to the outer coast than earlier Iron Age power centers and later towns like Bergen (Skre 2014).

Navigational practices before the diffusion of the magnetic compass are harder to reconstruct from the surviving evidence. However, it seems likely that the coastal environment of the Norðvegr and the long history of movement along this coast afforded a kind of low-instrumental navigation that was widespread by the time transatlantic voyages began to be conducted in the late ninth century. This does not mean that navigation was an unstructured, subjective process; indeed, experiments showing that it is possible to accurately estimate the time of day while at sea (Börjesson 2009; Engvig 2001; Bill, personal communication 2021), along with evidence for collectively held average speeds for different kinds of vessel (Indruszewski and Godal 2006: 24) and Morcken's (1978) proof of a common system of measurement existing from at least the twelfth century point to a highly developed but profoundly different tradition of navigation to our own.[2] The collective nature of navigation seems also to have been important throughout this period. In the *Saga of the Greenlanders*, Bjarni consults with his crew before setting off toward Greenland (Smiley and Kellogg 2001: 637), and cooperation between multiple crew members appears to be depicted on the Gotland picture stones and the Bayeux tapestry (Ellmers 1995).

---

2 Although the evidence presented by Morcken for the common use of *vikur* as distance units is convincing, his arguments for instrumental navigation in the Viking Age have been extensively criticized and, in some cases, disproven (Filipowiak 2020; Sayers 2003).

Finally, Viking Age and early medieval sailors also seem to have had a time-based and navocentric conception of their voyages and made collective judgments based on perceived risk. Adam of Bremen and the *Landnámabók* both report on sailing voyages using units of time rather than distance (Adam of Bremen 2002: 218–219; Pálsson and Edwards 2007: 16), and the *Landnámabók* author describes these voyages as a sequence of places one needs to sail past, as do the accounts of Ohthere and Wulfstan (Bately and Englert 2007; Englert and Trakadas 2009). The identification of landmarks relative to the direction of the ship's movement is apparent in both these accounts and occurs throughout Norse poetry (Jesch 2015). Geographical descriptions such as that of Greenland given to Bjarni Herjólfsson in the *Saga of the Greenlanders* are given from the perspective of the perceiver (Smiley and Kellogg 2001: 637), suggesting that Norse seafarers would not have conceived of their known world from a top-down perspective. Risk judgment is perhaps best exemplified in the *Historia de profectione Danorum in Hierosolymam,* in which survival at sea is directly attributed to the "ability to rightly judge the route and the sea" (Gertz 1922: 480; translation by Stephan Borgehammar, personal communication 2023).

This brief assemblage of evidence shows that the affordances identified during this project's trials have strong parallels with maritime movements and practices from the Viking and high medieval periods. Nevertheless, we should not take such parallels as indicators of environmental, cultural, or technological stasis among the coastal communities of the Norðvegr over the last 12 centuries (Bill 2010). Nor should we assume that similarities in movement and practice equate to similarities of experience or worldview (Meulengracht Sørensen 1995). The purposes, beliefs, and cultural backgrounds of the modern voyagers who were engaged in these trials remain undeniably distinct from those of a Viking Age or high medieval boat crew.

But as we have seen, practical engagement also resulted in changes among the modern crews regarding conceptions of maritime space and travel. This seems to have occurred through a kind of peer-pressure of practice: the commonly inhabited space of experience, the use of specific language and terminology, the particular and almost ritualistic ways of performing certain tasks, the repetition of geographically situated names and stories, and the constant evaluation of perceived risk encouraged the adoption of a wholly different way of thinking. I suggest that this ontological shift occurred because traditional attitudes and approaches were more useful than conventional modern ones for the activities and evaluations involved in traditional sailing. This is in line with Hutchins' (1995: 66) arguments that different navigational ontologies developed out of different contexts of use, and that it was their usefulness, not their accurate depiction of reality, that determined their adoption.

Characterizing an Ontology of Viking Age and High Medieval Seafaring

The establishment of an experiential nexus with ancient seafarers is apparent in the changes of perspective among the modern crews toward attitudes and practices evident in the archaeological and historical record. This connection through practice serves as a window not only on everyday activity but also on the broader maritime worldviews of the seafaring communities under study. We can therefore use the experimental and historical evidence assembled in this chapter to propose some possible characteristics of a Viking Age and high medieval seafaring ontology from the Norðvegr. Such an ontology seems to have consisted of the following: (1) an ever-changing set of named and storied places, used as mnemonic anchors across (2) a meshwork of possible routes, which together created (3) a boundless but thoroughly interconnected seascape. The shared conceptualization of this seascape was based not on universal representations of it but rather on (4) a dynamic and not wholly universal tradition of usage founded on practical experience in maritime activities. This practice-based geography led to the prevalence of (5) temporal or proportional systems of measurement founded on (6) commonly held averages rather than absolute values (average sailing and rowing speeds, average sizes of body parts used in boatbuilding), and a common understanding of the conditions under which such averages were applicable. This afforded (7) a navocentric system of orientation from which the world was taken in as a seamless integration of directly perceivable elements and what was believed to exist beyond. Voyages through this watery world involved many negotiations, but the central factor affording route choice seems to have been (8) a collective judgment of perceived risk to vessel and crew.

Conclusion: A Maritime Cultural Mindscape?

Ancient mobility, and particularly ancient maritime mobility, is an essential part of the human story that has proven difficult to analyze and represent with conventional historical and archaeological methods. The aim of this chapter has been to present an alternative approach that can bridge the ontological gap between modern researcher and ancient sailor. With this approach I have been able to identify the primary affordances of traditional Norwegian seafaring, evaluate their potential as a window on Viking Age and high medieval maritime ontologies, and suggest some of the possible characteristics of these ontologies. The result is a conditional, sequential, and relational seascape, with a voyage understood as a series of possible routes, and route choices determined primarily by collective judgments of perceived risk.

Over 30 years ago Christer Westerdahl identified an assemblage of maritime heritage that he used to define the maritime cultural landscape (Westerdahl 1989, 1992). The present study has revealed a conception of space and mobility that is equally embedded in coastal and maritime life and equally distinct from terrestrial worldviews. Hoping to move beyond the ideas of mental mapping and cognitive landscapes as direct representations of space in the brain, and inspired by Westerdahl's seminal work, I suggest that the ontology presented in this chapter is best understood as a maritime cultural mindscape. This mindscape is developed by a moving and perceiving person in the world and is therefore inherently contextual and alive; it spreads out from the perceiver to relate to other human and nonhuman agents; and it is not purely computational or representational but includes experience, skill, knowledge, and judgment, all of which play a role in actions and choices.

Various scholars within the Blue Humanities have suggested that approaches to ancient maritime activity could benefit from the inclusion of dynamism, contingency, and relationality (Campbell 2020; Steinberg and Peters 2015). The focus on experience and practice taken in this chapter, along with the concept of the maritime cultural mindscape, may serve to complement Westerdahl's work in this way. But beyond conceptual concoction, it is paramount to remember that the choices made by sailors in the past were founded on skill and wisdom, not data. The greatest challenge facing future models and analyses of ancient seafaring is, therefore, to base their representations and computations not only on quantitative environmental and nautical variables but also on culturally specific judgments of perceived risk and opportunity. As we have seen, both influenced ancient seafarers, as well as the nature and outcome of their voyages, and the wider, watery world through which they traveled.

Acknowledgments

The sailing voyages that make up the primary dataset for this project were never individual undertakings but were collective achievements. Without the brave, skilled, and resolutely positive individuals who participated in these trials, this project would not have been possible. Thanks go also to everyone at Fosen Folkehøgskole for sharing their wisdom and time and allowing me to collect data on board their boats. Finally, I would like to thank my supervisors Nicolò Dell'Unto and Jan Bill for their support and faith in my endeavors at sea.

Supplementary Data

The complete dataset collected during this project's sailing trials and trial voyages will be published as part of the author's doctoral thesis.

References Cited

Adam of Bremen. 2002. *History of the Archbishops of Hamburg-Bremen.* Translated by Francis J. Tschan. Records of Western Civilization. Columbia University Press, New York.

Bately, Janet, and Anton Englert (editors). 2007. *Ohthere's Voyages: A Late 9th-Century Account of Voyages along the Coasts of Norway and Denmark and Its Cultural Context.* Maritime Culture of the North 1. The Viking Ship Museum, Roskilde.

Berthelsen, Reidar. 1997. Kystfolket i jernalder og mellomalder. Fiskerbønder eller bondefiskere? In *Arkeologi og kystkultur,* edited by Helge Sørheim, pp. 6–15. Sunnmøre Museum, Ålesund.

Bill, Jan. 2010. Viking Age Ships and Seafaring in the West. In *Viking Trade and Settlement in Continental Western Europe,* edited by Iben Skibsted Klæsøe, pp. 19–42. Museum Tusculanum Press, Copenhagen.

Bischoff, Vibeke, Anton Englert, Søren Nielsen, and Morten Ravn. 2014. From Ship-Find to Sea-Going Reconstruction. In *Experiments Past: Histories of Experimental Archaeology,* edited by Jodi Reeves Flores and Roeland Paardekooper, pp. 233–47. Sidestone, Leiden.

Börjesson, Lena Lisdotter. 2009. Viking Age Navigation on Trial: Estimates of Course and Distance. In *Between the Seas: Transfer and Exchange in Nautical Technology: Proceedings of the Eleventh International Symposium on Boat and Ship Archaeology, Mainz 2006,* edited by Ronald Bockius, pp. 61–72. Verlag des Römisch-Germanischen Zentralmuseums, Mainz.

Brink, Stefan, and Neil Price (editors). *The Viking World.* Routledge, Abingdon, Oxon.

Broodbank, Cyprian. 2006. The Origins and Early Development of Mediterranean Maritime Activity. *Journal of Mediterranean Archaeology* 19(2): 199–230.

Campbell, Peter B. 2020. The Sea as a Hyperobject: Moving Beyond Maritime Cultural Landscapes. *Journal of Eastern Mediterranean Archaeology and Heritage Studies* 8(3–4): 207–225.

Chemero, Anthony. 2003. An Outline of a Theory of Affordances. *Ecological Psychology* 15(2): 181–195.

Crumlin-Pedersen, Ole, Erik Andersen, Bent Andersen, and Max Vinner. 1980. *Nordlandsbåden—analyseret og prøvesejlet af Vikingeskibshallens Bådelaug.* Nationalmuseet, Copenhagen.

Cunliffe, Barry W. 2017. *On the Ocean: The Mediterranean and the Atlantic from Prehistory to AD 1500.* Oxford University Press, Oxford.

David, Nicholas, and Carol Kramer. 2001. *Ethnoarchaeology in Action.* Cambridge World Archaeology. Cambridge University Press, Cambridge.

Edney, Matthew H. 1999. Reconsidering Enlightenment Geography and Map Making: Reconnaissance, Mapping, Archive. In *Geography and Enlightenment,* edited by David N. Livingstone and Charles W. J. Withers, pp. 165–198. University of Chicago Press, Chicago.

Eldjárn, Gunnar, and Jon Godal. 1988. *Nordlandsbåten of Åfjordsbåten Bind 1: båten i bruk: segling, roing, fisking og vedlikehald.* Dei gamle Forsto Mykje; 1. Kjelland, Lesja.

Eldjárn, Gunnar, and Jon Godal. 1990. *Nordlandsbåten og Åfjordsbåten Bind 4: system og oversyn.* Dei gamle Forsto Mykje; 1. Båtstikka, Rissa.

Ellmers, Detlev. 1995. Crew Structure on Board Scandinavian Vessels. In *Shipshape: Essays for Ole Crumlin-Pedersen,* pp. 231–40. Viking Ship Museum, Roskilde.

Englert, Anton. 2006. Trial Voyages as a Method of Experimental Archaeology: The Aspect of Speed. In *Connected by the Sea: Proceedings of the Tenth International Symposium on Boat and Ship Archaeology, Roskilde 2003,* edited by Lucy Katherine Blue, Frederick M. Hocker, and Anton Englert, pp. 35–42. Oxbow, Roskilde.

Englert, Anton. 2012. Travel Speed in the Viking Age: Results of Trial Voyages with Reconstructed Ship Finds. In *Between Continents: Proceedings of the Twelfth International Symposium on Boat and Ship Archaeology, Istanbul 2009,* edited by Nergis Günsenin, Søren Nielsen, and the International Symposium on Boat and Ship Archaeology, pp. 269–277. Yayinlari, Istanbul.

Englert, Anton, and Athena Trakadas (editors). 2009. *Wulfstan's Voyage: The Baltic Sea Region in the Early Viking Age as Seen from Shipboard.* Maritime Culture of the North 2. Viking Ship Museum, Roskilde.

Engvig, O. T. 2001. The Viking Way: Part 1. *Viking Heritage Magazine* 1: 3–7.

Færøyvik, Bernhard, and Oystein Færøyvik. 1979. *Inshore Craft of Norway.* Edited by Arne Emil Christensen. Conway Maritime Press, London.

Filipowiak, Wojciech. 2020. How Vikings Crossed the North Atlantic? The Reinterpretation of "Sun Compasses"—Narsarsuaq, Wolin, Truso. *International Journal of Nautical Archaeology* 49(2): 318–328.

Frog, Maths Bertell, and Kendra Willson. 2019. Introduction: Looking Across the Baltic Sea and Over Linguistic Fences. In *Contacts and Networks in the Baltic Sea Region: Austmarr as a Northern Mare nostrum, ca. 500–1500 AD.* Turku Medieval and Early Modern Studies, edited by Maths Bertell, Frog, and Kendra Willson, pp. 11–25. University of Amsterdam Press, Amsterdam.

Gertz, M. C. 1922. *Scriptores Minores Historiae Danicae Medii Aevi.* Vol. 2. G.E.C. Gad, Copenhagen.

Gibson, James J. 2014. *The Ecological Approach to Visual Perception: Classic Edition.* Psychology Press, New York.

Gillings, Mark. 2012. Landscape Phenomenology, GIS and the Role of Affordance. *Journal of Archaeological Method and Theory* 19(4): 601–611.

Hafting, Torkel, Marianne Fyhn, Sturla Molden, May-Britt Moser, and Edvard I. Moser. 2005. Microstructure of a Spatial Map in the Entorhinal Cortex. *Nature* 436(7052): 801–806.

Hasslöf, Olof, Henning Henningsen, and Arne Emil Christensen. 1972. *Ships and Shipyards, Sailors and Fishermen: Introduction to Maritime Ethnology.* Copenhagen University Press, Copenhagen.

Hutchins, Edwin. 1995. *Cognition in the Wild.* 1995. MIT Press, Cambridge, Massachusetts.

Indruszewski, George, and Jon Godal. 2006. Maritime Skills and Astronomic Knowledge in the Viking Age Baltic Sea / Pomorske spretnosti in zvezdoslovno védenje v vikinškem času na Baltiku. *Studia mythologica Slavica* 9: 15–39.

Indruszewski, George, Jon Bojer Godal, and Max Vinner. 2008. The Art of Sailing Like Wulfstan. In *Wulfstan's Voyage: The Baltic Sea Region in the Early Viking Age as Seen from Shipboard,* edited by Anton Englert and Athena Trakadas, pp. 274–293. Maritime Culture of the North 2. Viking Ship Museum, Roskilde.

Ingold, Tim. 2000. *The Perception of the Environment: Essays on Livelihood, Dwelling and Skill.* Routledge, London.

Ingold, Tim. 2011. *Being Alive: Essays on Movement, Knowledge and Description.* Routledge, Abingdon, Oxon.

Jarrett, Greer. 2025. From the Masthead to the Map: An Experimental and Digital Approach to Viking Age Seafaring Itineraries. *Journal of Archaeological Method and Theory* 32, art. 42, https://doi.org/10.1007/s10816-025-09708-6.

Jesch, Judith. 2015. *The Viking Diaspora.* The Medieval World. Abingdon, Oxon.

Jónsson, Guðni (editor). 1947. *Íslendinga sögur. Bd 10: Austfirðinga sögur.* Íslendingasagnaútgáfan, Reykjavík.

Kemp, John, and Brian D'Olier. 2016. Early Navigation in the North Sea: The Use of the Lead and Line and Other Navigation Methods. *Journal of Navigation* 69(4): 673–697.

Kruse, Arne. 2017. On Harbours and Havens: Maritime Strategies in Norway During the Viking Age. In *Viking Encounters: Proceedings of the 18th Viking Congress,* edited by Søren M. Sindbaek and Anne Pedersen, pp. 170–185. Aarhus University Press, Aarhus.

Lane, Paul. 2016. Editorial. *World Archaeology* 48 (5): 605–608.

Larson, Laurence Marcellus (translator). 1917. *The King's Mirror (Speculum Regale—Kunungs Skuggsjá).* The American-Scandinavian Foundation, New York.

Lisman, John, György Buzsáki, Howard Eichenbaum, Lynn Nadel, Charan Ranganath, and A. David Redish. 2017. Viewpoints: How the Hippocampus Contributes to Memory, Navigation and Cognition. *Nature Neuroscience* 20(11): 1434–1447.

Marcus, Geoffrey J. 1980. *The Conquest of the North Atlantic.* Boydell, Woodbridge, Suffolk.

Mathisen, Mariann, and Arne-Terje Sæther. 2018. *Nordlandsbåt og draug: En felles kulturarv.* Kasavi, Tromsø.

Meulengracht Sørensen, Preben. 1995. Ottars verdensbillede, religion og etik. *Ottar* 5:48–53.

Morcken, Roald. 1978. *Veien mot nord: Vikingetidens distansetabell langs den norske kyst fra svenskegrensen til Hvitehavet.* Sjöfartshistorisk Årbok. Bergens sjøfartsmuseum, Bergen.

Østmo, Einar. 2020. The History of the Norvegr 2000 BC–1000 AD. In *Rulership in 1st to 14th century Scandinavia,* edited by Dagfinn Skre, pp. 3–66. De Gruyter, Berlin.

Pálsson, Hermann, and Paul Edwards (translators). 2007. *The Book of Settlements: Landnámabók.* University of Manitoba Press, Winnipeg.

Parsons, Gavin. 2013. Gaelic Bards and Norwegian Rigs. *Journal of the North Atlantic* 4: 26–34.

Ravn, Morten. 2016. Om bord på vikingetidens langskibe—En analyse af besætningsor-ganisation og kommunikation. *Kuml* 65(65): 131–152.

Reyes-García, Victoria, Debora Zurro, Jorge Caro, and Marco Madella. 2017. Small-Scale Societies and Environmental Transformations: Coevolutionary Dynamics. *Ecology and Society* 22(1), https://www.jstor.org/stable/26270045.

Ruiz-Puerta, Emily J., Greer Jarrett, Morgan L. McCarthy, Shyong En Pan, Xénia Keighley, Magie Aiken, Giulia Zampirolo, et al. 2024. Greenland Norse Walrus Exploitation Deep into the Arctic. *Science Advances* 10(39), https://doi.org/10.1126/sciadv.adq4127.

Safadi, Crystal, and Fraser Sturt. 2019. The Warped Sea of Sailing: Maritime Topographies of Space and Time for the Bronze Age Eastern Mediterranean. *Journal of Archaeological Science* 103: 1–15.

Sayers, William. 2003. Karlsefni's "húsasnotra": The Divestment of Vinland. *Scandinavian Studies* 75(3): 341–350.

Skre, Dagfinn. 2014. Norðvegr–Norway: From Sailing Route to Kingdom. *European Review* 22(1): 34–44.

Smiley, Jane, and Robert Kellogg. 2001. *The Sagas of the Icelanders: A Selection.* Penguin, New York.

Steinberg, Philip, and Kimberley Peters. 2015. Wet Ontologies, Fluid Spaces: Giving Depth to Volume Through Oceanic Thinking. *Environment and Planning D: Society and Space* 33(2): 247–264.

Storli, Inger. 2007. Ohthere and His World. A Contemporary Perspective. In *Ohthere's Voyages: A Late 9th-Century Account of Voyages along the Coasts of Norway and Denmark and Its Cultural Context,* edited by Janet Bately and Anton Englert, pp. 76–99. Maritime Culture of the North 1. The Viking Ship Museum, Roskilde.

Sturlason, Snorre. 1990. *Heimskringla, or The Lives of the Norse Kings.* Edited by Erling Monsen. Translated by Albert Hugh Smith. Dover, New York.

Tolman, Edward C. 1948. Cognitive Maps in Rats and Men. *Psychological Review* 55(4): 189.

Turville-Petre, G., and E. S. Olszewska, trans. 1942. *The Life of Gudmund the Good, Bishop of Holar.* Curtis & Beamish, Coventry.

Valtonen, Irmeli. 2008. The North in the "Old English Orosius." A Geographical Narrative in Context. *Neuphilologische Mitteilungen* 109(3): 380–384.

Wernke, Steven A., Lauren E. Kohut, and Abel Traslaviña. 2017. A GIS of Affordances: Movement and Visibility at a Planned Colonial Town in Highland Peru. *Journal of Archaeological Science* 84 (August): 22–39.

Weski, Timm. 2006. The Value of Experimental Archaeology for Reconstructing Ancient Seafaring. In *Connected by the Sea: Proceedings of the Tenth International Symposium on Boat and Ship Archaeology, Denmark 2003,* edited by Lucy Katherine Blue, Frederick M. Hocker, and Anton Englert, pp. 63–67. Oxbow, Roskilde.

Westerdahl, Christer. 1989. *Norrlandsleden 1 Källor till det maritima kulturlandskapet: En handbok i marinarkeologisk inventering = Sources of the maritime cultural landscape: A handbok of marine archaeological survey.* Länsmuseet Murberget, Härnösand.

Westerdahl, Christer. 1992. The Maritime Cultural Landscape. *International Journal of Nautical Archaeology* 21(1): 5–14.

Westerdahl, Christer. 2010. Ancient Sea Marks: A Social History from a North European Perspective. *Deutsches Schiffahrtsarchiv* 33: 71–155.

Whittington, James C. R., David McCaffary, Jacob J. W. Bakermans, and Timothy E. J. Behrens. 2022. How to Build a Cognitive Map. *Nature Neuroscience* 25(10): 1257–1272.

13

Negotiating Watery Worlds

Crafting a Research Agenda

Colin Grier, Mikael Fauvelle,
and Alberto García-Piquer

However one measures it, our planet includes more than one million kilometers of coastline. Since time immemorial, humans have dwelled upon, sailed from, and derived sustenance from this extensive and unique element of our planet. The same can be said for other kinds of watery worlds—rivers, lakes, wetlands, and, indeed, the open ocean. Our past, present, and future are intimately connected with the 72% of the Earth's surface that is covered in water. The Pacific Ocean alone covers roughly one-third of the Earth. As such, water has profoundly shaped our histories, our practices, our ways of being in the world, and, as Jordi Rivera Prince points out in Chapter 11 of this volume, our very bodies themselves.

As the breadth of chapters in this volume make clear, there is a great diversity of ways humans have engaged with these watery worlds, and a similarly diverse set of approaches can be pursued to account for and interpret these experiences. In this final chapter we argue that to achieve a rich understanding of seafaring practices, we must embrace and document this diversity of engagements. But, as we describe below, we must simultaneously push for an integrated and holistic approach to the study of seafaring that highlights common themes, practices, and histories in relation to how we as a species have negotiated watery worlds.

Below we outline what we see as the foundations of a diversity-embracing yet expansive, holistic, and globally comparative seafaring analysis project. This chapter involves three components. First, we amplify some of the many threads and themes explored in this volume as reflective of the diversity of strategies and approaches to seafaring that humans used in the past. Second, we illustrate several ways we can approach seafaring so that the study of "going by boat" moves beyond tweaking terrestrial models to fit ocean circumstances. Here we venture into the arena of theory-building—that is, launching an effort to build out theory from an initial seafaring perspective, homing in on ways in which watercraft are transformative rather than simply additive to human practices and potential. Third, we lay out a synthetic research agenda that will, in our view, take our collective seafaring project forward in important and potentially transformative ways in relation to archaeological, anthropological, and interdisciplinary scholarship. Our hope is that this endeavor and the directions charted will spur scholars to carry on this project in new ways that lead to transformative understandings of seafaring in our human past.

Appreciating Diversity in Seafaring Practices

The component of this planet that is or that interfaces with water is incredibly diverse. From ice-choked arctic oceans to far-flung tropical archipelagos to rugged fjordlands, the complexity of coastal and ocean environments is astonishing (Figure 13.1). Layering on to this complexity are the diversity of societies that have inhabited these waters, further multiplying the range of contexts we must consider under the umbrella of seafaring.

In focusing this volume, we have skimmed over the larger polities and monumental watercraft to amplify the arena that we feel has been less explored—that of small-scale, coastal seafaring peoples. We want to bring these societies into the larger discussion, despite the sometimes significant lack of direct material evidence pertaining to seafaring in their archaeological records. This volume has been an effort to bring them into a more central focus in the stories and histories we construct for seafaring peoples. Narratives pertaining to the seafaring practices of small-scale peoples often relate to exploring new worlds and to the origin point of coastal adaptations and seafaring technologies. But there is much more to tell. Ethnographic data help in this aim, as does modeling, as several chapters in our volume (e.g., García-Piquer: Chapter 2; Rorabaugh: Chapter 3) attest and argue.

True to the theme of recognizing diversity, we also must appreciate that non-Western and pre-state societies had quite different strategies for negotiating the Earth's watery worlds than ships operating in the context of large-scale polities. Of course, small-scale societies had more modest seafaring craft and used them not typically for regular cross-ocean voyages but for negotiating a complex set of coastlines in the world of daily practices. Traversing coastlines and, indeed, river systems and smaller freshwater bodies often has travelers within sight of land and able to make a quick maneuver toward shore. This creates a different context and field of play for boat use in smaller-scale societies. This to not to

Figure 13.1. The diversity of global coastlines, illustrated by (*clockwise from top left*) the southern Gulf Islands of coastal British Columbia, the fjordlands of Norway, the southern California coast, and the coastline of Patagonia in South America.

ignore that modest craft were certainly used in times and places for expansive ocean-traversing journeys, such as we see with the Lapita expansion and the colonization of Polynesia (Furholt et al. 2020). Indeed, peoples in small-scale societies made some large and impressive watercrafts. But as Kenneth Ames (2002) points out, in small-scale societies it is important to recognize boats not as an exceptional technology used under exceptional circumstances but as a central and often quotidian instruments of daily production and practices.

To grapple with this, we can consider boats as sitting at the nexus of a set of social relations of production, including those that pertain to the production of watercraft themselves. The production of a large galleon is a different game than the production of almost any craft used by small-scale societies. Peoples of the Northwest Pacific Coast, for example, did construct exceptionally large war canoes, and these required a significant contingent of people to build and propel them. And at the same time, they generated family-sized watercraft for basic resource gathering needs. The organization of their production was specific to the craft and the social context, and qualitative and quantitative differences between this and large commercial shipping (for example) certainly exist in the social relations of their production. This is itself worthy of study, as we elaborate below.

That said, "small-scale seafaring" is not a single strategy or approach, and we again must be careful to recognize the diversity of craft that were mobilized even within the same context in small-scale societies. As Mikael Fauvelle and Peter Jordan (this volume: Chapter 8) clearly show, different boats (tule versus dugouts) were produced through very different means for very different purposes within coastal California itself, with significant implications for the political realm and the centralization of power. Recognizing, appreciating, and analyzing diversity in practices is—as always—a matter of scale and context. As such, one key pillar of a productive seafaring project is to fully account for and appreciate the diversity of practices as a multiscalar endeavor.

Reformulating Theory

While appreciating the diversity of seafaring practices across time and space is important, it is equally important to pursue theory-building to generate fresh ideas and larger interpretive frameworks for seafaring as a central component of human practices and histories globally. We see a range of arenas in which watercraft have been transformative, as outlined in our introductory chapter. These include (1) providing access to new resources, (2) allowing for exploration and colonization of new regions, (3) facilitating the transportation of goods and people, (4) underwriting settlement networks and mobility strategies, (5) expanding networks for social and biological reproduction, (6) shaping social relationships and political strategies, and (7) generating ontologies of seafaring (Knapp 2020).

For our purposes here, we find it useful to fold these into three broader arenas in which we can pursue theory-building, and which we ultimately argue constitute key domains in the holistic, integrated, and global framework we wish to advance. These arenas include (1) logistics and movement, (2) sociopolitical organization and change, and (3) ontologies and phenomenology. Below we cover these three arenas of theory-building separately, then follow this up with an outline of the integrated framework we feel can carry the study of seafaring forward.

Logistics and Movement

Logistics and movement have traditionally been where archaeologists and other scholars of seafaring have tried to build general theory. This has typically included considerations of the technological capacities of watercraft and the impacts of widespread movement on the spatial organization of small-scale societies. Collecting and foraging models as outlined by Lewis Binford (1980) have long been used to try to capture variation and patterning in terrestrial hunter-gatherer movements, to reasonable effect. Binford always argued that collectors and foragers represent not "types" of strategies but a continuum, and those who utilize coasts and oceans—the so-called maritime hunter-gatherers—have been viewed as representing an extreme on the collecting end (Ames 2002; Binford 1990; Yesner 1980).

At the same time, there has been recognition that maritime hunter-gatherers may be qualitatively different from their terrestrial counterparts (e.g., Suttles 1968; Yesner 1980). Boat travel is not simply a faster type of walking or analogous to riding horses, as Matthew Des Lauriers and Claudia García-Des Lauriers (this volume: Chapter 5) point out. Some have undertaken the project of trying to build a middle-range theory (that is, a generalizing account) of maritime adaptations and maritime hunter-gatherers, notably David Yesner (1980) in his effort to capture their key organizational elements in a set of ten organizational principles. Yet such synthetic approaches reveal a long-standing limitation of this type of theory-building—that maritime hunter-gatherers are a coherent, definable, and distinct variant of what we understand terrestrial hunter-gatherers to be. Interestingly, watercraft garner almost no concerted attention in Yesner's (1980) paper, except that they facilitate greater movement and thus promote a more "logistically oriented" form of settlement system.

How do we get beyond the quantitative differences and effectively capture the qualitative differences that seafaring engenders? What approaches best represent taking seafaring on its own terms? Ames (2002) offered a starting point for this, building from Binfordian foraging models by considering the impact of boat mobility on a wide range of productive pursuits. Yet the expansion of a foraging range is itself transformative, not just additive. A circle's area increases exponentially relative to its radius, so incorporating a greater area can also incorporate a greater diversity of resources and other people (Bicho and Esteves 2022). An expanded territory is not just about distance covered on water but the structure of resources and other people that fall inside it as well as the way seafaring and watercraft allow humans to "map onto" that ecological structure and social field in novel ways not typical of terrestrial approaches.

Several examples can be cited to illustrate how we might rethink seeing the world when it incorporates water and seafaring. First, a study by Michael Blake (2010) of movements in the central Salish Sea of western coastal North America illustrates a critical point about the complexity of movements and territories when watercraft are involved. In analyzing the structure of Sechelt territories, Blake argues that the extent of their territories was defined by the distance one could travel in a boat plus an additional distance inland or upland to the extent that a trek to the edge of a territory and back is possible in two days. Boats and terrestrial conveyance are thus not mutually exclusive;

they are often used in combination. Indeed, that combination seems to be a predominant driver in defining the extent and organization of First Nations territories in the coastal inlets of southwestern British Columbia. Hybridization of movements rather than either/or is key here. As Blake (2010) notes, this kind of hybridized movement strategy has a significant structuring effect on not only the extent of territories but the distribution of settlements within those territories, creating archaeological patterns that may not be readily predicted by terrestrial foraging theory.

A second case in point emerges from a study by Farid Rahemtulla (2006) in which he examines lithic procurement and the organization for production of lithic technologies at the long-inhabited Namu village on the central coast of British Columbia. Rahemtulla argues that, based on their lithic assemblages, the people at Namu must have had watercraft technology underpinning their lithic procurement over at least the last 5,000 years. As tool stone is heavy and often sourced in remote locations (at least in coastal British Columbia), the acquisition of a consistent supply of tool stone can require significant time and energy. Rahemtulla (2006) argues that those requirements are mitigated when embedded procurement strategies are used in which multiple resources are acquired during wide-ranging forays using watercraft. The problems of lithic procurement are resolved in the context of other problems. As such, seafaring makes humans consummate multitaskers and makes the nature of trips multipurpose and complexly layered.

An implication of Rahemtulla's study, and the notion of embedded procurement generally, is that predictions concerning transport pathways and mobility strategies from the perspective of single resources are perhaps less appropriate for seafaring peoples than for territorial groups who cover less ground. While terrestrial collectors solve resource conflicts through logistical mobility, embedded procurement is the simultaneous resolution of several problems in an integrated spatial fashion. Analogically, it may be akin to the argument made by Gregory Monks (1987) about Northwest Pacific Coast resource procurement— that Northwest coast hunter-gatherers exploit whole food chains not individual resources, and we must evaluate their decision-making in that frame. Another implication of Rahemtulla's study is that it concerns watercraft use at a village that was permanently occupied in its location for millennia (Cannon 2003), thus offering an approach and perspective useful for modeling seafaring agricultural communities that are as firmly planted in their settlement locations.

A final example of how we might build an approach to movement from a seafaring vantage point is provided by the study in this volume by Alberto García-Piquer (Chapter 2). Through sophisticated modeling of seas and currents in southern South America, his work effectively illustrates how various seas and waterways may facilitate or hinder coastal passage. To explore and understand the sea, it is best to treat it as a complex environment rather than a consistent friction-bearing surface and include portaging as a key element of seafaring. Different boats also interact differently with the dynamics of the sea. As a result, we must model the coastal and open-ocean environments as seafarers see and experience them. While this can be and often is done for terrestrial models by building in aspects of terrain and ground cover to least-cost path approaches, getting beyond the sea as essentially a flat plane of constant movement costs is an important arena of both method and theory-building (more on this below). Moreover, while water is often viewed from a terrestrial perspective as a barrier, it can be a barrier to or facilitator of travel varyingly at different times, depending on a set of circumstances that, unlike terrestrial environments, are constantly changing in essentially real time. These examples point to starting with a seafaring-driven view of the structure of the watery worlds we wish to understand.

Sociopolitical Organization and Change

To paraphrase Rhoda Halperin (1989), boats do not build themselves, nor do they propel themselves or move resources themselves—rather, it is humans who do all those things, and those humans have relations among them that govern the mobilization of this technology in myriad ways. Adopting this stance shifts our study of watercraft from the arena of what Halperin called "locational movements"—that is, the spatial structure of movements on a landscape or seascape—into the realm of "appropriational movements," which involves the analysis of how people and their relationships are organized with respect to deploying resources and technologies.

We feel this is the logical extension of what Ken Ames (2002) was arguing for in his seminal paper, which can be construed in Marxian terms as focusing on the "social relations of production." Who builds boats? Who controls or has access to the materials and knowledge to build boats? Who has access to the use of boats, and under what pretenses and circumstances? These are all questions that usefully begin without the assumption that if there are boats, then everyone has and uses boats. As archaeologists have long engaged in the discussion of political economies (Furholt et al. 2020), the political and power-imbued component of boat use seems inescapable given their transformative capacities. With this, the social organization of watercraft production and seafaring sociopolitics becomes a central domain in a larger seafaring project (Faar 2006).

Case studies that underscore this point have appeared in the wider literature and are represented in this volume as well. The analysis of the emergent inequalities that stem from the whaling enterprise among ethnographically and archaeologically documented whaling communities in the North American Arctic offers an object lesson in how boat ownership is central to the reorganization of sociopolitical systems in precontact Arctic whaling communities (Burch 1981; Cassell 1988; Grier 2000; Whitridge 2002). Similarly, on the Northwest Pacific Coast of North America, smaller boats were ubiquitous, but the scale of labor organization required to operate larger boats often involved factions, lineages, or corporate households (Ames 2002; Erin Smith, this volume: Chapter 9). Canoe construction was usually the craft of an elite person trained in vessel making, with some assistance by family and other less trained labor. Elites by birthright possessed the history, knowledge, and spirit power to mount the production of a vessel and ensure its safe journeying. Thus, elites controlled many aspects of boat production. As such, they were instruments of inequality, and the complexity of the ways they operated in the political arena requires careful investigation.

In this volume, we are treated to several studies that take on this political economy of watercraft directly. First, Fauvelle and Jordan adeptly point out that the production of complex watercraft may have initially resulted from efforts to solve logistical issues. But these craft provided new opportunities and potentialities that fueled political change toward political hierarchies and centralization. Fauvelle and Jordan (Chapter 8) note that this is not the necessary direction that political change must move but that it is evident in a significant number of case studies of early and small-scale maritime complexity. As such, their comparative work represents a useful example of how we can build general theory out of usefully controlled and relevant comparisons.

Building on this point that complex watercraft do not inherently result in hierarchies or centralization, Victor Thompson (Chapter 7) emphasizes collective action as a component of how complex watercraft systems were implemented. The Calusa built extensive canal systems for watercraft that were collectively rather than centrally managed, providing a mechanism for integration and the construction of collective social institutions that operated at local scales. Recognizing such systems, and the social practices that govern them, expands our understanding of not just the complexity of watercraft systems but also of political systems themselves. This places watercraft at the center of developing alternatives to the construction of political hierarchies.

In this vein, we might productively engage the idea that watercraft can offer the means to decentralize sociopolitical practices, a theme that has emerged more widely in the discussion of resistance to centralization and anarchism approaches to understanding political histories and change (e.g., Angelbeck and Grier 2012; Furholt et al. 2020; Graeber and Wengrow 2021). Several case studies provided by Martin Furholt and colleagues (2020) describe how boats are used as one mechanism to evade, escape, or otherwise thwart increasing social control and political hierarchies. Those pushed to margins of political control or those actively resisting efforts to centralize authority can use boats to subvert authority and break the "bottlenecks" that are critical for political centralization to emerge.

In this sense, the use of watercraft expands our understanding of political dynamics and how three varying trajectories of change—centralization, collective action, and decentralizing forces—work separately and in concert in the context of the political histories of smaller-scale political systems. As Thompson (this volume: Chapter 7) points out, the emergence of watercraft systems as an integrating and collective action-driven process may down the road fuel the emergence of much larger-scale and centralized political entities. Perhaps the most significant element of watercraft across this range of processes is that they provide for greater interaction over larger networks and can therefore support fundamentally different kinds of interactive networks than we have presumed work in terrestrial contexts.

We can thus move the study of watercraft and seafaring to a central role in studying the emergence of various kinds of social and political networks and can see watercraft as instruments of political change—a key arena we can productively explore and from which we may generate useful new theory. In this pursuit it is critical to recognize that all technologies are implemented (or not) in a specific social context. The reasons watercraft are used initially may not be how they later get used, nor are certain watercraft technologies always used when they can be. When they become useful, they are mobilized, with "becoming useful" meaning politically rather than solely logistically useful.

Ontology and Phenomenology

As we have argued, watercraft are more than just a technology and a means of transporting people and goods. Another key arena in which this is true is in the realm of what watercraft mean to seafarers. In support of this we can consider the most basic and obvious analogy—the automobile in the twentieth century. America in the second half of the twentieth century became effectively a car-driven culture. The technology facilitated and supported a new way of life, a way of thinking about what is possible in society, a source of identity construction individually and collectively, and an expression of the affluence and prowess of American industrial know-how. In these and many other respects cars have and continue to express and embody an American way of being in the world that is based in many ways around the automobile.

The analogy with watercraft is simplistic perhaps, but nonetheless instructive, as Smith (this volume: Chapter 9) points out by using the term "watercraft culture." Seafaring the world over has been rife with meaning, and notions and practices generated by life at sea have worked their way into many aspects of meaning in daily life, such as the naming of boats being similarly applied to cars, vernacular expressions of human behaviors such as "running a tight ship," and various other sea-based metaphors and translations that are drawn from the watery worlds we have collectively and individually engaged.

As such, we need to account for the many meanings and ontologies of watercraft locally and collectively, as several papers in this volume do to great effect (e.g., Jarrett, Chapter 12; Smith, Chapter 9; Whitridge, Chapter 10). What do watercraft mean to those who use them, and how are they part of relating to and representing our relationship with watery worlds? How do such perceptions structure the ways in which people interact with the world? This ontological arena represents one in which we may productively pursue theory-building.

For this we can tap into the broader theoretical canon of posthumanism (e.g., Cipolla et al. 2021; Crellin et al. 2021; Crellin and Harris 2020) and its ability to frame how we relate to the world in more profound ways than as a problem to solve (Golledge 2003). On the water, there are relentlessly practical problems to solve—assessing currents, monitoring water depth, timing a landing, and so on. These can often be issues of survival and thus bring humans to terms with their rather frail existence subject to forces that are difficult to negotiate, let alone control or master. The sea thus presents a perfect environment in which our complex relations with the world—material and nonmaterial—are poignantly realized.

The posthumanist focus on relational ontologies and heterarchy of agency among actors (human and nonhuman) in the world thus seems ideally suited to pursue both the diversity of and shared human experiences on the sea. Adopting the posthumanist canon of theory may be a matter of effective borrowing and translating but, as we illustrated for sociopolitical approaches, is also a matter of expansion. Smith (this volume: Chapter 9), for example, adopts the notion of homologies to bridge and fuse the land/sea divide ontologically. Peter Whitridge (this volume: Chapter 10) takes a similar stance in pursuing an account of Inuit–animal–thing relations. His adoption of actor-network theory grounds the mental and physical in the material record itself, reinforcing that actor relations are folded in many ways into the material archaeological record of seafaring that may seem impoverished because of the lack of archaeological boats. Indeed, the attention that actor-network theory calls to the material expressions of these relationships represents a fruitful (yet nonetheless demanding) basis for an archaeology of seafaring, one that takes us well beyond the traditional places we may think to look for our evidence.

A key piece we would add to this ontological approach is that of phenomenology (Johnson 2012; Tilley 1994), in order to ground seafaring in the physicality of experiencing watery worlds. Seafaring is simultaneously mental (cognitive), experiential (historical), and somatic (sensory and immediate). Anyone who has spent time on the water and in smaller craft (below the size and scale of ocean-going vessels) realizes this. Negotiating water is about many things simultaneously—knowledge and experience, confidence on the water, the technical conditions of the craft and crew, and one's savvy for adapting to ever-changing conditions. We have all been humbled by the raw energy of oceans and coasts, even large lakes or current-heavy rivers. Such intimate experiences are individual as well as shared and collective but nonetheless are mediated through the physicality of being on the water. The key point of all this is that seafaring is a physical act, and the commonality of the way we all experience the sea physically can form the basis for addressing our shared experiences with it, a point we feel extends from the goals of a phenomeneological approach in archaeology (Johnson 2012).

Our physical relationship with the sea has a historical dimension as well. As Rivera Prince points out in Chapter 11, the effects of watery worlds on us physically are not just momentary but rather imprint on our very bodies, with the sea shaping our physical being. The sea is also reshaping our future as well, as Indigenous communities are reconstituting their traditional practices, relations, and ontologies through reconnections with the sea, something alienated from them in the colonial past (Smith, this volume: Chapter 9). In this way the sea offers another dimension of restorative justice, an ontology and world of meaning that can be reclaimed in pursuit of a more equitable future.

Toward an Integrated Approach

In many respects this chapter represents not a conclusion but a starting point. Each of the chapters in this volume has opened up important conversations that need to be taken forward. Our goal in this final chapter has been to illuminate and weave together a bigger picture, first by promoting an appreciation of the diversity of seafaring practices, and then by using that breadth to work toward more generalizing theory and approaches. While appreciating diversity and pursuing generalization have at times been at odds in the production of knowledge, we see both as essential components of knowledge production, meshed inextricably in a dialectic full of productive tensions and synergistic moments.

The final aim we have in this chapter is to lay out the groundwork for an integrated approach to seafaring and outline in an explicit way the components that we feel cover the bases required to properly conceptualize and study seafaring practices. With this we hope not just to capture maritime adaptations, the history of boat use, seafaring as wayfinding, and all the other elements of negotiating watery worlds that have been raised in this volume but to comprehend seafaring as a relational existence in an immersive context. Below we map out the architecture of an integrated approach to studying seafaring as we conceive of it.

The studies presented in this volume, and those that it hopefully spurs, can be usefully seen as positioned in relation to our three major themes—logistics and movement, sociopolitical organization and change, and ontology and phenomenology. Appropriating perhaps the most quintessentially terrestrial model, we can use the "soil triangle" for terrestrial sediment description—appropriately altered for our seafaring purposes—to convey the range of inquiry covered in our integrated approach (Figure 13.2). The key point we wish to make with Figure 13.2 is that productive studies of seafaring practices, while perhaps situated toward or emphasizing one of the vertices of the triangle, explicitly incorporate the other dimensions of studying seafaring and pay explicit attention to each of these poles in whatever mix seems appropriate to the study. With this schema, there is, for example, no useful logistics and movement chapter that fails to recognize that logistics and movement are inextricably connected with the sociopolitical and the relational.

The approach that knits this volume together is that each of the papers herein in some way takes that on mission—to weave together elements of the three poles rather than being singularly concerned with one of them. This requires focused effort. Indeed, in wrestling with the traditional approach to organizing an edited volume in a way that delineates sections and arranges papers in some linear or logical sequence, we recognized quickly that the chapters are related in a multifaceted rather than linear way. Nor are they best organized geographically, since the themes they address transcend regions. In this sense the diversity-amplifying and generalizing nature of the papers provide an instance of how we envision an integrated approach to seafaring unfold.

The Role of Modeling

One component we wish to highlight that may provide an important building block in our approach is modeling. This volume is not heavily focused on modeling, and none of the three editors would define ourselves as modelers explicitly. Yet an approach to seafaring that involves modeling—that is, uses simulations and agent-based structured decision-making—is, in our minds,

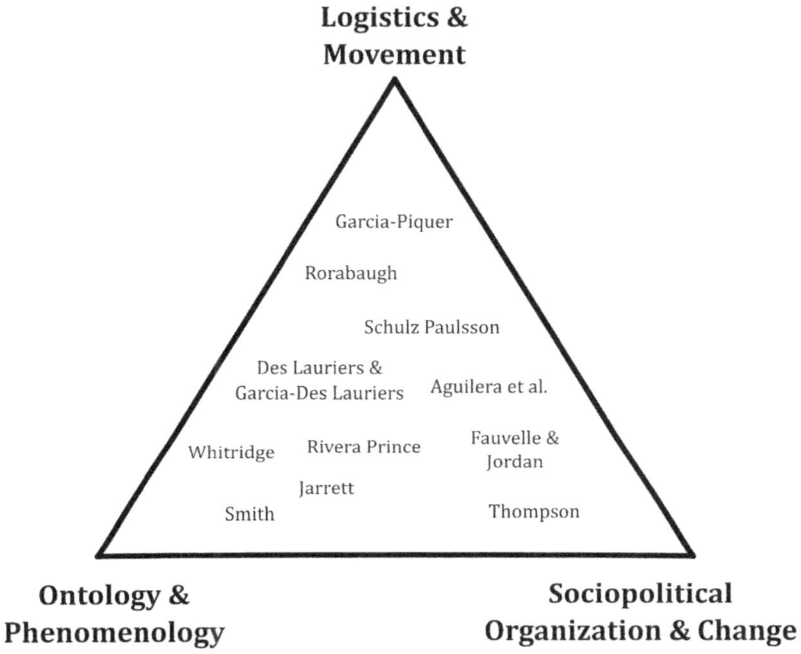

Logistics & Movement

Garcia-Piquer

Rorabaugh

Schulz Paulsson

Des Lauriers & Garcia-Des Lauriers   Aguilera et al.

Whitridge   Rivera Prince   Fauvelle & Jordan

Jarrett

Smith   Thompson

Ontology & Phenomenology   **Sociopolitical Organization & Change**

Figure 13.2. Schematic triangle illustrating the themes (triangle vertices) that form our integrated approach, with individual papers in this volume positioned accordingly.

a critical element of moving forward with our seafaring project, for several reasons.

First, given the current dearth of archaeological data on boats themselves, models can help evaluate the relative technological capacities of various watercraft systems under various conditions. This is perhaps the most obvious application of modeling, and elements of it are found formally and more informally in several chapters in this volume, such as those by Alberto García-Piquer (2), Greer Jarrett (12), and Adam Rorabaugh (3).

Second, modeling provides a framework to assemble and evaluate the variables that are relevant to seafaring. Modeling also helps to explicitly stipulate relevant sea conditions and evaluate the impacts of varying conditions on boats given a specific technology. Despite many coastal archaeologists doing fieldwork in boats, we often have only a limited understanding of the conditions and circumstances related to negotiating the sea. Modeling allows us to incorporate additional variables and evaluate their effects, such as shifting shorelines and water level fluctuations, to craft nuanced simulations that build upon but go beyond any direct experience (see García-Piquer, this volume: Chapter 2). In this light, experimental archaeology emerges as an invaluable complement to modeling. Jarrett (this volume: Chapter 12) makes a cogent case that there is no substitute for real experience on the water, but integrating direct experience with modeling provides a framework to evaluate a range of hypothetical possibilities that are simply not possible to evaluate out on the water.

Third, we can go beyond the boats and sea conditions themselves to model networks of people and how boats make possible, or whether they uniquely make possible, certain kinds of social networks. This realm has been underexplored, but social network analysis in a GIS frame set in a modeling environment holds potential to both simulate the emergence and stability of networks and test hypotheses concerning the role boats can and have played in supporting various kinds of networks, as is well demonstrated by the contribution from García-Piquer in this volume (Chapter 2).

A final point on modeling is that it is possible in the way we structure our models to build in the ontologies of seafaring we glean from ethnographic or archaeological data. Modeling need not be a distanced, objective, or generalizing tool. Indeed, chapters in this volume that take a modeling stance (García-Piquer [2], Rorabaugh [3], and, in a more experimental sense, Jarrett [12]) all attempt to integrate some way of being in the world into their approach. Can we push such approaches further, evaluating the influence perceptions of the sea and ideational maps of the world have on archaeological patterns?

The Way Forward

Seafaring has always been a concern of archaeologists and those who recognize that our world is primarily one of water. In archaeology alone, there is clearly an abundance of literature on various aspects of seafaring, much of which we have only been able to draw in tangentially, if at all. For instance, several recent edited volumes in the British Archaeological Reports Series "Maritime Archaeology for the 21st Century" bring together multiple methodological and theoretical dimensions of seafaring, including *Delivering the Deep* (Ilves et al. 2024) and *Down by Water* (Vadillo et al. 2022). These collections of works highlight a diversity of aspects of seafaring and often adopt a productive interdisciplinary approach. But they stick mostly to the subject of logistics and movement of people and goods in more centralized and complex societies. The volume *Marine Ventures* (Bjerck 2016) offers an expansive set of chapters, with several venturing into the realm of sociopolitical organization as we define it here. Beyond these edited volume efforts (naming just a few), the *Journal of Island and Coastal Archaeology* has provided a venue to explore and aggregate coastal and island studies, and many themes we raise in this volume have also been productively explored therein. Surveying the existing literature, the components and pieces are there but have yet to be drawn into an overarching framework.

Much of this literature is in English; here, for practical purposes, we limit our coverage to scholarly literature primarily in English. Part of our call for an integrated approach must include drawing together information from as many languages and scholarly traditions as we can to address local/regional diversity and engage in productive theory-building. We also need to deeply mine the ethnographic record—much of it in languages other than English—to illuminate non-Western perspectives and ontologies of seafaring, a goal productively demonstrated by the chapter offered herein by Nelson Aguilera and colleagues (Chapter 6).

In this volume we have been less concerned with the traditional epic transoceanic journeys and discovery of new continents than with the daily practices that emerge with seafaring and the mechanisms that shape and transform those practices in coastal areas where land and sea mesh. We have made an explicit effort to illuminate the rhythm of daily practices involving watercraft, going beyond watercraft as a technology, beyond seafaring as a mode of production, and ultimately entering into and conceptualizing a realm where seafaring represents an immersive context. Over 20 years ago, Ken Ames' (2002) notion of watercraft as instruments of daily production provided a stepping-stone and key impetus to engage such an approach, but we still have a lot of water to cover.

References Cited

Ames, Kenneth M. 2002. Going by Boat: The Forager-Collector Continuum at Sea. In *Beyond Foraging and Collecting: Evolutionary Change in Hunter-Gatherer Settlement Systems,* edited by Ben Fitzhugh and Junko Habu, pp. 19–52. Kluwer Academic / Plenum, New York.

Angelbeck, Bill, and Colin Grier. 2012. Anarchism and the Archaeology of Anarchic Societies: Resistance to Centralization in the Coast Salish Region of the Pacific Northwest Coast. *Current Anthropology* 53(5): 547–587.

Bicho, Nuno, and Eduardo Esteves. 2022. Pleistocene Hunter-Gatherer Coastal Adaptations in Atlantic Iberia. *Frontiers in Earth Science* 10: 957214.

Binford, Lewis R. 1980. Willow Smoke and Dogs' Tails: Hunter-Gatherer Settlement Systems and Archaeological Site Formation. *American Antiquity* 45(1): 4–20.

Binford, Lewis R. 1990. Mobility, Housing, and Environment: A Comparative Study. *Journal of Anthropological Research* 46: 119–152.

Bjerck, Hein Bjartmann (editor). 2016. *Marine Ventures: Archaeological Perspectives on Human–Sea Relations.* Equinox, Sheffield.

Blake, Michael. 2010. Navegación y definición de territorios en la Costa Noroeste de Norteamérica: Un ejemplo Coast Salish. In *La excepción y la norma: Las sociedades indígenas de la Costa Noroeste de Norteamérica desde la arqueología,* edited by A. Vila and J. Estévez, pp. 84–93. Treballs d'Etnoarqueologia No. 8, CSIC-U.A.B., Barcelona.

Burch, Ernest S. 1981. *The Traditional Eskimo Hunters of Point Hope Alaska, 1800–1875.* The North Slope Borough, Barrow.

Cannon, Aubrey. 2003. Long-Term Continuity in Central Northwest Coast Settlement Patterns. In *Archaeology of Coastal British Columbia: Essays in Honour of Philip M. Hobler,* edited by Roy L. Carlson, pp. 1–12. Archaeology Press, Simon Fraser University, Burnaby, B.C.

Cassell, Mark S. 1988. Farmers of the Northern Ice: Relations of Production in the Traditional North Alaskan Inupiat Whale Hunt. *Research in Economic Anthropology* 10: 89–116.

Cipolla, Craig N., Rachel J. Crellin, and J. T. Harris. 2021. Posthuman Archaeologies, Archaeological Posthumanisms. *Journal of Posthumanism* 1(1): 5–21.

Crellin, Rachel J., Craig N. Cipolla, Lindsay M. Montgomery, Oliver J. T. Harris, and Sophie V. Moore. 2021. *Archaeological Theory in Dialogue: Situating Relationality, Ontology, Posthumanism and Indigenous Paradigms.* Routledge, London.

Crellin, Rachel J., and Oliver J. T. Harris. 2020. What Difference Does Posthumanism Make? *Cambridge Archaeological Journal* 31: 469–475.

Faar, Helen. 2006. Seafaring as Social Action. *Journal of Maritime Archaeology* 1: 85–99.

Furholt, Martin, Colin Grier, Matthew Spriggs, and Timothy Earle. 2020. Political Economy in the Archaeology of Emergent Complexity: A Synthesis of Bottom-Up and Top-Down Approaches. *Journal of Archaeological Method and Theory* 27: 157–191.

Golledge, Reginald. 2003. Human Wayfinding and Cognitive Maps. In *Colonization of Unfamiliar Landscapes: The Archaeology of Adaptation,* edited by Marcy Rockman and James Steele, pp. 25–43. Routledge, London.

Graeber, David, and David Wengrow. 2021. *The Dawn of Everything: A New History of Humanity.* Penguin, London.

Grier, Colin. 2000. Labor Organization and Social Hierarchies in North American Arctic Whaling Societies. In *Hierarchies in Action, Cui Bono?,* edited by Michael W. Diehl, pp. 264–283. Center for Archaeological Investigations Occasional Paper No. 27, Southern Illinois University, Carbondale.

Halperin, Rhoda. 1989. Ecological Versus Economic Anthropology: Changing "Place" Versus Changing "Hands." *Research in Economic Anthropology* 11: 15–41.

Ilves, Kristin, Veronica Walker Vadillo, and Katerina Velentza (editors). 2024. Delivering the Deep: Maritime Archaeology for the 21st Century. Selected Papers from IKUWA 7. BAR International Series 3170.

Johnson, Matthew H. 2012. Phenomenological Approaches in Landscape Archaeology. *Annual Review of Anthropology* 41: 269–284.

Knapp, A. Bernard. 2020. Maritime Narratives of Prehistoric Cyprus: Seafaring as Everyday Practice. *Journal of Maritime Archaeology* 15: 415–450.

Monks, Gregory G. 1987. Prey as Bait: The Deep Bay Example. *Canadian Journal of Archaeology* 11: 119–142.

Rahemtulla, Farid. 2006. Design of Stone Tool Technology During the Early Period (10,000–5,000 BP) at Namu, Central Coast of British Columbia. PhD dissertation, Simon Fraser University.

Suttles, Wayne. 1968. Coping with Abundance: Subsistence on the Northwest Coast. In *Man the Hunter,* edited by Richard B. Lee and Irvin DeVore, pp. 56–68. Aldine, Oxford.

Tilley, Christopher. 1994. *A Phenomenology of Landscape: Places, Paths and Monuments.* Berg, Oxford.

Vadillo, Veronica Walker, Emilia Mataix Ferrándiz, and Elisabeth Holmqvist (editors). 2022. *Down by the Water: Interdisciplinary Studies in Human–Environment Interactions in Watery Spaces.* Bar International Series 3108.

Whitridge, Peter. 2002. Social and Ritual Determinants of Whale Bone Transport at a Classic Thule Winter Site in the Canadian Arctic. *International Journal of Osteoarchaeology* 12: 65–75.

Yesner, David R. 1980. Maritime Hunter–Gatherers: Ecology and Prehistory. *Current Anthropology* 21: 727–750.

CONTRIBUTORS

Nelson Aguilera Águila received his PhD in prehistoric archaeology from the Autonomous University of Barcelona, Spain.

Matthew R. Des Lauriers is associate professor of anthropology at California State University, San Bernardino. He is the author of *Island of Fogs*.

Mikael Fauvelle is an associate professor and researcher in the Department of Archaeology and Ancient History at Lund University. He is the coeditor of *An Archaeology of Abundance: Reevaluating the Marginality of California's Islands*.

Claudia García-Des Lauriers is associate professor of anthropology at California State Polytechnic University, Pomona. She is the coeditor of *Archaeology and Identity on the Pacific Coast and Southern Highlands of Mesoamerica*.

Alberto García-Piquer is a postdoctoral researcher in the Department of Prehistory at the Autonomous University of Barcelona. He is the coeditor of *Beyond War: Archaeological Approaches to Violence*.

Colin Grier is professor of anthropology at Washington State University. He is the coeditor of *Beyond Affluent Foragers: Rethinking Hunter-Gatherer Complexity*.

Greer Jarrett is a PhD candidate in the Department of Archaeology and Ancient History at Lund University.

Peter Jordan is a professor in the Department of Archaeology and Ancient History at Lund University. He is the author of *Technology as Human Social Tradition: Cultural Transmission Among Hunter-Gatherers*.

Bettina Schulz Paulsson is currently a researcher and lecturer at the University of Gothenburg. She is the author of *Time and Stone: The Emergence and Development of Megaliths and Megalithic Societies in Europe*.

Raquel Piqué is professor of archaeology at the University of Barcelona. She is the author of *Arqueología del Hain: Investigaciones etnoarqueológicas en un sitio ceremonial de la sociedad selknam de Tierra del Fuego.*

Jordi A. Rivera Prince is a Presidential Postdoctoral Fellow in the Department of Anthropology at Brown University.

Adam N. Rorabaugh is adjunct faculty in the Department of Archaeology at Simon Fraser University.

Erin M. Smith is an adjunct professor in the Department of History, Anthropology, Modern Languages and Literatures at Eastern Washington University.

Victor D. Thompson is a Distinguished Research Professor and the executive director of the Georgia Museum of Natural History at the University of Georgia. He is the coeditor of *The Archaeology of Villages in Eastern North America* and *The Archaeology and Historical Ecology of Small Scale Economies.*

Peter Whitridge is professor of archaeology at Memorial University of Newfoundland.

INDEX

Society and Ecology in Island and Coastal Archaeology
Edited by Victor D. Thompson and Scott M. Fitzpatrick

The settlement and occupation of islands, coastlines, and archipelagoes can be traced deep into the human past. From the voyaging and seafaring peoples of Oceania to the Mesolithic fisher-hunter-gatherers of coastal Ireland, to coastal salt production among Maya traders, the range of variation found in these societies over time is boundless. Yet, they share a commonality that links them all together—their dependence upon seas, coasts, and estuaries for life and prosperity. Thus, in all these cultures there is a fundamental link between society and the ecology of islands and coasts. Books in this series explore the nature of humanity's relationship to these environments from a global perspective. Topics in this series range from edited volumes to single case studies covering the archaeology of initial migrations, seafaring, insularity, trade, societal complexity and collapse, early village life, aquaculture, and historical ecology, among others along islands and coasts.